The International Legal Governance of the Human Genome

The human genome is a well-known symbol of scientific and technological progress in the twenty-first century. However, concerns about the exacerbation of inequalities between the rich and the poor, the developing and the developed states, and the healthy and the unhealthy are causing problems for the progress of scientific research. International organisations such as UNESCO, the Human Rights Council, the European Union, the World Health Organization and the General Assembly have engaged in developing international law to regulate biotechnology in the last two decades.

Human rights are central to the debate on the regulation of research on the human genome. A human rights approach to the governance of human genomic research would be piecemeal and ineffective so long as fundamental issues about economic, social and cultural rights, the so-called second generation of human rights, are not addressed. This book argues that, in order to be able to meaningfully apply a human rights framework to the governance of the human genome, the international human rights framework should be based on a unified theory of human rights where the distinction between positive and negative rights is set aside.

In analysing the Universal Declaration on the Human Genome and Human Rights (adopted by the General Conference of UNESCO in 1997), the book argues for a common heritage concept with the right to development at its core and explores the content of the right to development through rational human rights theory. The Universal Declaration on Bioethics and Human Rights (2005) supports such a conception of the framework laid out by the 1997 Universal Declaration.

Central to concerns about addressing inequalities in health research are the notions of property embedded in intellectual property regimes that govern such research at the national level. It is argued that the notion of property rights in the human genome should be placed within the context of protecting human rights, including the right to development. The concept of common heritage of humanity, contrary to the widely held understanding that it is in opposition to patenting of gene sequences, supports human rights-based conceptions of property rights.

This book fills a gap in the literature on international legal governance of the human genome and will provide an essential reference point for research into the right to development, development issues in bioethics, the role of international institutions in law making and research governance.

Chamundeeswari Kuppuswamy is a lecturer in the Law School of the University of Shefffield, UK.

Genetics and Society

Series Editors: Paul Atkinson, Associate Director of CESAGen, Cardiff University; Ruth Chadwick, Director of CESAGen, Lancaster University; Peter Glasner, Professorial Research Fellow for CESAGen at Cardiff University; and Brian Wynne, member of the management team at CESAGen, Lancaster University.

The books in this series, all based on original research, explore the social, economic and ethical consequences of the new genetic sciences. The series is based in the ESRC's Centre for Economic and Social Aspects of Genomics, the largest UK investment in social-science research on the implications of these innovations. With a mix of research monographs, edited collections, textbooks and a major new handbook, the series will be a major contribution to the social analysis of new agricultural and biomedical technologies. Series titles include:

Governing the Transatlantic Conflict over Agricultural Biotechnology
Contending coalitions, trade liberalisation and standard setting
Joseph Murphy and Les Levidow

New Genetics, New Social Formations
Peter Glasner, Paul Atkinson and Helen Greenslade

New Genetics, New Identities
Paul Atkinson, Peter Glasner and Helen Greenslade

The GM Debate
Risk, politics and public engagement
Tom Horlick-Jones, John Walls, Gene Rowe, Nick Pidgeon, Wouter Poortinga, Graham Murdock and Tim O'Riordan

Growth Cultures
Life sciences and economic development
Philip Cooke

Human Cloning in the Media
Joan Haran, Jenny Kitzinger, Maureen McNeil and Kate O'Riordan

Local Cells, Global Science
Embryonic stem cell research in India
Aditya Bharadwaj and Peter Glasner

Handbook of Genetics and Society
Paul Atkinson, Peter Glasner and Margaret Lock

The International Legal Governance of the Human Genome
Chamundeeswari Kuppuswamy

Debating Human Genetics
Contemporary issues in public policy and ethics
Alexandra Plows

Community Genetics
Eugenics and genetic utopias/dystopia
Avid Raz

Genetic Testing
Accounts of autonomy, responsibility and blame
Michael Arribas-Ayllon, Srikant Sarangi and Angus Clarke

Scientific, Clinical and Commercial Development of the Stem Cell
From radiobiology to regenerative medicine
Alison Kraft

Genetically Modified Crops on Trial
Opening up alternative futures of Euro-agriculture
Les Levidow

The Making of a Syndrome
The case of Rett Syndrome
Katie Featherstone and Paul Atkinson

The International Legal Governance of the Human Genome

Chamundeeswari Kuppuswamy

Routledge
Taylor & Francis Group
LONDON AND NEW YORK

First published 2009
by Routledge
2 Park Square, Milton Park, Abingdon, Oxon, OX14 4RN

Simultaneously published in the USA and Canada
by Routledge
270 Madison Avenue, New York, NY 10016

Routledge is an imprint of the Taylor & Francis Group, an informa business

Typeset in Times New Roman by Pindar NZ, Auckland, New Zealand
Printed and bound in Great Britain by the MPG Books Group, Bodmin and King's Lynn

British Library Cataloguing in Publication Data
A catalogue record for this book is available from the British Library

Library of Congress Cataloging-in-Publication Data
Kuppuswamy, Chamundeeswari.
The international legal governance of the human genome / Chamundeeswari Kuppuswamy.
p. cm.
Includes bibliographical references.
1. Human genetics—Law and legislation. 2. Genetic resources—Law and legislation. 3. Human genetics—Moral and ethical aspects. I. Title.
K3611.A77K87 2009
344.04'196—dc22 2008050953

ISBN10: 0-415-45857-9 (hbk)
ISBN10: 0-203-92940-3 (ebk)

ISBN13: 978-0-415-45857-3 (hbk)
ISBN13: 978-0-203-92940-7 (ebk)

For Annapoorna, Amma and Daddy

Contents

Foreword

In recent times, international law, and in particular international human rights law, has had to develop innovative frameworks to deal with the scientific discoveries of our age: one such is the sequencing of the human genome. Several international organisations, principally UNESCO, WHO, the new Human Rights Council and the General Assembly have had to grapple with the legal issues raised by the progress of scientific research. These bodies have developed new law relating to biotechnology; similar advances have occurred at the European level under the auspices of the EU.

The application of international human rights law norms and, in particular, economic, social and cultural rights are central to the regulation of research on the human genome. This monograph breaks new ground in developing a human rights framework to the governance of the human genome. The author analyses several core documents: in particular, the Universal Declaration on the Human Genome and Human Rights 1998 and the Universal Declaration of Bioethics and Human Rights are given detailed, systematic and critical treatment. The author applies the concept of the 'common heritage of mankind' with the right to development at its core in her careful and balanced exposition of this fundamental element of her thesis. This study deals also with critical issues relating to intellectual property regimes governing national research.

The monograph makes a significant contribution to the whole complex of issues relating to the legal governance of the human genome. The author places her subject carefully in the context of general international law, and her research makes a landmark and ground breaking contribution to the literature on the subject.

Sandy Ghandhi
University of Reading
3 November 2008

Preface

When I first started working on the project of examining how international law was responding to new technology, I focussed on the topic in a rather narrow sense. I intended to scrutinise the application of the concept of the common heritage of mankind (hitherto used, for instance, in relation to the resources of the deep seabed) in the field of genomics. I soon realised that this was a vastly different project to mere application; it was going to lead me to revisit fundamental notions in international law. I have ventured into philosophy, bioethics, genetics and international relations. The foundations of my understanding on various themes in this book come from adopting an inter-disciplinary approach. The focus on the legal aspects in parts and on non-legal areas in others also stems from this interdisciplinary approach which led me to believe that legalistic discussions in some areas might be premature, and attempting predominant legal interpretations of certain topics would not be just and equitable. So I have tried to focus on extra-legal narratives, particularly in the area of intellectual property. The chapter on the right to development is detailed in order to focus on the crucial nature of the right, and to distinguish it from other rights in the ever growing list of human rights. The final version has concentrated on providing a panoramic picture of the global governance of the human genome – the international organisations involved, the history of the issues and concerns that have arisen at the global level in connection to new technology. It is intended for a wide range of scholars and policy makers. Without delving into the details, I can conclude that the research for this book has left me with numerous topics to follow up on, such as the legality of the Nuremberg Code, the nature of institutional acts of Specialised Agencies, their competences and the legislative process, some which have been completed but not included in this book, others left to fill another manuscript.

In summary this book argues that an international legal approach to the human genome should be based on a human rights framework underpinned by a unified theory of human rights and that property and development should be addressed from such a perspective.

Chamu Kuppuswamy
29 October 2008

Acknowledgements

I thank Professor Nigel White, Professor Sandy Ghandhi and Dr Duncan French for reading the chapters and giving me comments. I thank Professor Deryck Beyleveld for introducing me to Alan Gewirth's theory and for his own unique style of helping me assimilate various aspects of the theory. I am indebted to participants at various conferences and congresses who sat through and commented on the ideas that went into this book. I would like to thank colleagues and staff at Sheffield Law School. I am grateful to staff at the British Library and the IALS library where I carried out a lot of my research. I thank all colleagues and staff at the United Nations University, Institute of Advanced Studies (UNU-IAS), Tokyo where I was a PhD Fellow between 2002–03. The time I spent at UNU-IAS broadened my horizons by providing me the opportunity to meet and interact with representatives and delegates from various countries, to understand the complexities involved in interstate affairs and the diffuse nature of international governance. My heartfelt thanks to Professor Darryl Macer (formerly University of Tsukuba) who gave me much support and encouragement throughout my time at Tokyo.

This book is primarily based on research conducted for the completion of my PhD, which I received in 2007. I finally wish to thank my friends at Sheffield, particularly in the early years of my study in England for having provided me with moral support and friendship. All errors are completely mine.

Chamu Kuppuswamy
29 October 2008

Table of cases

A&M Records, Inc. v. Napster, 114 F. Supp. 2d 896 (ND Cal. 2000).

Case Concerning the Barcelona Traction, Light and Power Company Limited, Belgium v. Spain, 1970, *ICJ* 3.

Case No. 9, The Nuremberg Trials: The Einsatzgruppen Case, 29 September 1947–9 April 9 1948, Military Tribunal II.

Diamond v. Chakrabarty, 447 US 503, 1980.

Government of South Africa v. Grootboom and Others, Constitutional Court – CCT11/00 2001 (1) SA 46 (CC), 4 October 2000.

Moore v. Regents of the University Of California, 793 P. 2d 479 (Cal. 1990).

Pharmaceutical Manufacturer's Association of South Africa and Others and the President of the Republic of South Africa, The Honourable Mr N. R. Mandela N. O. and Others v. Treatment Action Campaign, *Amicus Curiae* Case No. 4183/98 (Transvaal Provincial Division).

Texaco Overseas Petroleum Company/California Asiatic Oil Company v. the Government of the Libyan Arab Republic (Arbitration), *ILM* XVII, 1–37.

Thomas F. Deuel, Yue-Sheng Li, Ned R. Siegel and Peter G. Milner, *Re*, No. 94–1202, United States Court of Appeals, Federal Circuit, March 28, 1995, 51 *Federal Reporter, Third Series*, 1552.

United States v. Brandt, United States v. Flich, Trials of War Criminals before the Nuremberg Military Tribunals under Control Council Law No. 10. Nuremberg, October 1946–April 1949, Washington, DC: US GPO, 1949–1953.

United States v. Josef Alstötter, *et al.* ('The Justice Case'), 3 *TWC* 1, 1948, 6 *LRTWC* 1, 1948, 14 *Ann. Dig.* 278, 1948.

'I will give you a talisman. Whenever you are in doubt, or when the self becomes too much with you, apply the following test. Recall the face of the poorest and the weakest man whom you may have seen, and ask yourself, if the step you contemplate is going to be of any use to him .... Will it restore him to a control over his own life and destiny? In other words, will it lead to swaraj [freedom] for the hungry and spiritually starving millions? Then you will find your doubts and your self melt away.'

Mohandas Karamchand Gandhi (2 October 1869–30 January 1948), cited in Pyarelal, *Mahatma Gandhi: The Last Phase*, Ahmedabad, India: Navajivan Press, 1958, vol. 2, p. 65.

1 The human genome and bioethics

1.1 Introduction

Advances in science and technology have enabled us to alter the human person. It is now possible to replace organs, change one's appearance drastically, extend one's life by long periods and even change one's gender. There is very little we cannot tinker with, in the human body and the natural world. To this long list of achievements, we can now add the discovery of the 'book of life'.[1] The Human Genome Project (HGP) has revealed, more than any other biology project, the nature of the most important molecules of life – genes. Genes are considered the 'Holy Grail' of biology,[2] and have unlocked the potential for a quantum leap in our understanding of the human body, the functions within it and of life itself.[3] The human genome is a well-known symbol of scientific and technological progress. There is also no end to the rhetoric surrounding the human genome. This 'secular equivalent of the soul'[4] is the sum total of the genes in the human body and determines a lot of what we are and will become. Since the completion of the HGP in 2003, other projects have come to the forefront. Through the Human Proteome Project, stem-cell research, nanotechnology, robotics and artificial intelligence, we can expect to see the acquisition of more capabilities that are potentially species-altering in nature.[5]

In 1952, James Watson, Francis Crick, Rosalind Franklin and Maurice Wilkins discovered the double helical structure of the DNA.[6] Three of these scientists received the Nobel Prize for their discovery. It revolutionised the understanding of the fundamentals of biology by shifting the focus from the protein molecules, as a basic molecule, to the DNA as a building block of life. In 1953, the first commercially successful general-purpose computer was built by IBM (International Business Machines), as part of the Korean War effort.[7] Thomas Watson, Jr, in his quest for a 'defence calculator' to aid in the policing of Korea by the United Nations (UN), developed the prototype of the home computer. Only 19 of these machines were produced, 13 of which went to the US Department of Defense and the defence industry. Both these developments, along with other scientific breakthroughs, are having a revolutionary impact on biological research and in turn impact health research and health care. As a result, the traditional actors in medicine have also expanded and now include academic researchers and technologists. As more and more research is being funded by public and private bodies whose primary purpose for funding such research is to benefit the taxpayers and share holders, it is expected that such scientific research address issues of direct short-term importance to individuals and groups as well as benefit people in the long term.

1.2 The Human Genome Project (HGP)

The objective of the HGP was to determine the complete sequence of the three billion DNA subunits (technically known as bases – A, T, G, C), identify all human genes, and make them accessible for further biological study.[8] Scientists embarked on this project because of its fundamental nature. The project aroused immense interest because 'The particular order of As, Ts, Cs, and Gs is extremely important. The order underlies all of life's diversity, even dictating whether an organism is human or another species such as yeast, rice, or fruit fly.'[9]

The HGP symbolises the summit of technological excellence in biology at the turn of the century, paving the way for a big role for molecular genetics in the twenty-first century. It has spawned a whole new set of scientific laboratories and has given birth to a new generation of scientists with a fresh outlook on biology. It is a flagship molecular genetics project of the late twentieth century that has been instrumental in thrusting molecular genetics into the international limelight and has made it a matter of political concern at the highest level of interstate relations.

Japan was the first country to start a human genome project, as early as in 1981, when the Science and Technology Agency of Japan (STA), a government department, funded a project on the extraction, analysis, and synthesis of DNA. The project focused on DNA sequencing and involved technology heavyweights such as Seiko, Hitachi and others who produced the earliest DNA sequencing machines, micro-chemical robots, electrophoresis gel systems and other high-tech products, which are essential for analysing DNA. The Japanese project was extended into a second phase in 1984, funded by the STA, under the title 'Generic Basic Technologies to Support Cancer Research', making an early link between genomics and health benefits.[10] It later died away as the US played down the competition it faced from Japan in trying to develop high-technology products.

Around the mid-1980s, technologists and mathematicians, particularly in the US, became excited with the idea of designing methods for mapping the units in DNA molecules found in the nucleus of the human cell. They put their efforts into the sequencing of the units present in the DNA and successfully completed sequencing smaller organisms with shorter DNA. The idea that the same could be done with human DNA was taking root in the minds of technologists. The length of the DNA and the complexity involved in mapping the genome caused scepticism among some biologists while others even found it boring. But to computer engineers it posed challenging algorithmic questions.[11]

Those who opposed the funding of the project did not think it would be worth pursuing the project and accused the supporters of advocating 'big science'[12] and raised fears about the spread of genetic determinism whereby humans are considered the sum total of their genes. This is also known as reductionism, and the fear is that human activities would be reduced to a mere function of their genes. There were suspicions that the HGP would bring in eugenics by the backdoor.[13] Geneticists however claimed that it would be very useful to obtain a sequence map of the DNA, which could then be used to identify individual genes. Earlier, the gene for Huntingdon's disease had been identified using other methods, and was hailed as a major breakthrough in medicine.[14] Eventually geneticists were able to convince the majority about the immense importance of DNA sequencing and its huge potential to overcome human disease and to maintain and enhance human health. It was emphasised that such a project would be of interest to other specialists and in addressing human disease. The promise that this project showed was held to be massive. Even the common cold, referred to as a multifactorial disease (caused by a combination of factors: genetic, environmental and lifestyle factors), could benefit from the knowledge derived from the HGP.[15] Considering the changing nature of scientific research, with the applications of such research to health being a major part of the case put forward for the funding of such research nowadays, it is inevitable that there is scrutiny of how and whether such promises have been kept.

Aided by a favourable set of circumstances and the involvement of charismatic personalities,[16] the international HGP commenced in 1990, mainly in laboratories across the US funded by the Department of Energy and the Department of Health. Other participating countries were Italy, the UK, the USSR, France, Germany, Denmark, the EC, Canada and Japan. The main contribution to the sequencing of the genome came from the UK and the US, which completed about 75 per cent of the map. As the project went along, many innovations in international collaborations were made, thus laying the foundations of scientific standard-setting in the new area of genomics. The Human Genome Organisation (HUGO) was formed in 1988 to bring together the scientists working on the sequencing of the human genome.[17] Informal rules for sharing the rough sequences were made (the 'Bermuda Rules') so that labs across the world could access information, almost in real time,[18] which demonstrated that the HGP was a scientific project that was ground breaking in more than one way. Yet another innovation was the parallel research into the ethical and legal implications arising from the mapping of the genes.

Genomics was born into the era of speed, reflecting the nature of the world since the last decades of the twentieth century. The signature statement of the iconic head of the project, Jim Watson, was very much writ into it: 'Choose a scientific goal and push towards it relentlessly.'[19] Not only was the human genome being sequenced through the public consortium involving international collaborators, but a private firm, Celera Genomics, was independently involved in decoding the human genome.[20] The HGP was completed in 2003, two years ahead of schedule, when the whole sequence of the human DNA containing three billion subunits was unveiled to the world by President Clinton and Prime Minister Blair as the 'book of life', accompanied by much fanfare and many expectations for the future.[21]

The HGP has heralded a new age in biology and holds the promise to revolutionise medicine. New journals, intended to publish research from projects that use molecular-genetics methods, have sprouted and grown in strength.[22] High-speed computing and bioinformatics (use of mathematics and computer technology in biological and medical research) are equal and indispensable partners in genomics research. The technologists have developed new skills and produced new equipment for ever more accurate data collection. Analysts predict that 'in the coming years, humanity will come to use the computer more and more as a "substitute mind" – or language – to manipulate, redirect, and organise the vast genetic information that makes up the physical substance of living nature.' Private venture capital funding has changed the culture of scientific research in medicine. The HGP has had direct impact on medical practice, not only in the area of diagnosing hereditary diseases, but more broadly in the diagnosis of other diseases. The impact of the HGP is best described as 'making inroads into the medical mind', to borrow the phrase of an American specialist.[23] Personalised medicine is a phrase that is entering into medical practice, referring to the tailoring of treatment according to the genetic make-up of the individual. No doubt, many look at 'the future of medicine (as) genetic'.

1.3 Scientific research: increasing inequities and inequalities

Concerns about the exacerbation of the inequalities between the rich and the poor, the developing and developed States, and the healthy and the unhealthy are issues that are fundamental to the governance of scientific research. With so much being predicted by way of linking the HGP to health benefits, concerns about the exacerbation of inequalities in health are not unfounded. The fear that genomic science will lead to widening of the gap between the healthy and the unhealthy is not unwarranted. The scepticism of commentators that the progress of genomic science would indeed not help in addressing health issues relevant to majority of the global population is strengthened in the light of analysing the development of internationally applicable rules in the regulation of medical research using humans or materials derived from humans since the Second World War. There are hardly any systematic attempts to address the issues of inequalities and inequities in health amongst various populations in the world. Neither is there any sustained attempt to deal with this matter at an international level. The Nuremberg Code provided the first breakthrough for international law to consider such issues by promoting ethical reflection on the conduct of scientific experiments on human subjects and the rights of such subjects. But the post-Nuremberg period was a missed opportunity to develop international bioethical law. The attempt to assert physician/scientist autonomy in research combined with the politics of the Cold War successfully kept the area of ethics on human experimentation and the burgeoning field of bioethics within the national realm, with light-touch regulation at the national level. Although the era of the Cold War is over, the quest for independence in research continues to be an issue in science, with research ethics committees, for example, being considered an obstacle to cross, for both the right and the wrong reasons.

The Nuremberg Code could have become what the Universal Declaration of Human Rights is to human rights – an early inspiration to further international engagement in the field. In fact, human rights and bioethics should really have developed side by side since the Second World War. Instead, bioethics faded out of the international realm while human rights law benefited largely from the Nuremberg Code, and the status of the human person was firmly established as a subject in international law. Although the central principle of informed consent in the Nuremberg Code does not directly address the issues that are currently dominating the international scene – inequalities in health and health care – it is important to examine the document as an international effort to establish rules that had as its aim the welfare and wellbeing of individuals both within and outside of sovereign territory. The subsequent section analyses the Nuremberg Code and how it came to be.

1.3.1 The Nazi Doctors' Trial (the 'Medical Trial')

A major undertaking of the victors of the Second World War was the establishment of the Nuremberg War Criminals Tribunal for Major War Criminals (IMT) and other subsequent tribunals to try crimes that had been allegedly committed by the captured war criminals, during the Holocaust. The Nuremberg Charter and the judgements of the IMT revolutionised international law and brought about numerous changes in its wake.[24] The proceedings at the IMT exposed the nature and extent of the atrocities that had taken place in Europe and the judgements of the tribunal laid the foundations for international criminal law,[25] and made advances in the laws of war, human rights, humanitarian laws and bioethics.

The Holocaust included inhumane medical experiments that were carried out by doctors in Germany with the permission and encouragement of the State regime. These experiments, all of which were conducted without the consent of the individuals concerned, took place during the years 1939 and 1944 in four phases.[26] During the trial, issues of ethical behaviour in conducting experiments invariably surfaced. 'Activist' judges such as Sebring had at the back of their mind the implications of the trial for future medical research. Judge Sebring's probing questions about the ethics of research and the concept of ethically based clinical research laid the foundations for what was to later become the Nuremberg Code.[27] American and British medical experts testified as to the 'scientific worth' of the experiments conducted during the Nazi regime – on the value of the so-called 'Nazi medicine'. Considerable discussion and exchange of views took place as to culpability for medical experiments and a 'code of practice' for medical researchers.[28] And out of it emerged the Nuremberg Code.

1.3.2 The Nuremberg Code[29]

One of the earliest bioethical documents in international law, often cited by researchers, research ethics committees, national and international policy-making bodies, is the Nuremberg Code. The Nuremberg Code is a set of ten ethical principles (termed 'points'), centred on the requirement of consent for experimentation on humans. [30] It forms part of the judgement of the IMT in the Nazi Doctors' Trial, decided in 1947.[31]

Initially, a set of six principles, recommended by the experts who testified at the Trial, the judges of the IMT expanded it to ten, on the basis that medical experiments should conform to 'certain basic principles ... in order to satisfy moral, ethical and legal concepts'.[32] They declared that human experiments that did not satisfy these conditions 'were contrary to "the principles of the law of nations as they result from the usages established among civilized peoples, from the laws of humanity, and from the dictates of public conscience."'[33]

The Code stipulates that experiments on human subjects should be conducted after obtaining voluntary informed consent from those participating. Such experiments should be the least hazardous to the individual, for the good of society and should be carried out by qualified and responsible persons. The onus is on the person who initiates, directs and engages in the experiment to obtain consent and ensure that it is voluntary and informed. The Code warns against impunity in delegation of the responsibility in obtaining informed consent. It embodies other principles, which are echoed in modern municipal legislation and decisions of ethics review boards of hospitals and universities. The indispensability of the experiment is a necessary condition for its conduct. Alongside this absolute necessity for the experiment, whose results cannot be obtained through any other experiment, it should also be for the 'good of the society'.

The guidelines on the design of the experiment, stipulated in point 3 of the Code, are similar to the stages of drug trials that are conducted by pharmaceutical companies, requiring animal experimentation and research into the disease causing organism prior to human experimentation. The welfare of the subject is put at the heart of the Code when it prescribes that care should be taken to 'avoid all unnecessary physical and mental suffering and injury' for the human subject and back-up facilities should be made available to cover for eventualities. Risky experiments could be conducted, provided the risk taken was proportionate to the benefit it was designed to provide. It stipulated against random, unnecessary experiments, and experiments which might, by chance, lead to death or disabling injury. The Code provided for the continuous operation of consent throughout the experimental process, whereby the individual could withdraw consent at any point in time.

The following principles can be gathered from the Nuremberg Code:

1 Informed Consent should be obtained.
2 No harm to be caused to the human subject.
3 Society must benefit from such experiments.
4 Experiments should be performed by qualified personnel.
5 Experiments should be conducted according to certain minimum standards.

In responding to the case at hand the judges laid out rules that apply to experiments on the human subject, but also gave their statements room to be interpreted in broader ways. On the third point listed above, it can be construed that not only was the intention to protect individuals from whimsical ideas of scientists and researchers, but also to create a positive duty to benefit society with their work. Given the nature of the immediate political urgencies at hand, it would not have been possible to dwell on such ideas, but the seed for further growth and understanding of it was sown in the Nuremberg Code.

1.4 International bioethical law after the Nuremberg Code

While the list of human rights and humanitarian law documents mentioned the Nuremberg Code in its Preamble and also incorporated in it is long, the international community missed a valuable opportunity to capitalise on the lead provided by the Code, by failing to assert the need for international governance of the burgeoning field of bioethics and medical ethics. The Nuremberg Code was the first comprehensive bioethical document that laid down general principles on human experimentation. Although not followed up by international legal rules, it was the first of a series of rules and regulations by self-governing bodies and national governments relating to human experimentations,[34] which is the mainstay of bioethics. Clinical bioethics, involving patient-doctor relationship, is a newer development in bioethics. Another important feature of early bioethical discourse is the engagement of the public in these issues, which resonates with the current situation in Europe and elsewhere, where public opinion is key to decision making in laws relating to genomics. The Nuremberg Nazi Doctors' Trial led to widespread public interest in the ethics of research through discussions in newspapers on whether or not to publish the Nazi doctors' notes containing 'valuable' medical information.

The United Nations Educational, Scientific and Cultural Organization (UNESCO) avoided the issue of human experiments and did not send a representative to the Nazi Doctors' Trial. Julian Huxley, the first Director-General of UNESCO, wrote of the purposes of UNESCO as being the promotion of 'the international application of science to human welfare'.[35] Its constitution, which proclaimed the purpose of the organisation to be one of steering science towards contributing to peace and security, expressly stated that peace would fail it were not based on the moral solidarity of mankind. Yet it declined to get involved in the international scientific evaluation of Nazi 'medicine'.[36] The use of medicine during the Second World War by the Nazis was portrayed as gross violations of human rights perpetrated by a totalitarian State regime with the help of a few perverted members of the medical profession. Therefore it was argued that the overwhelming majority of the profession conducted themselves true to the Hippocratic Oath[37] and there was no need for monitoring and oversight by independent agencies. The idea was that 'The Nuremberg Code was a good code for barbarians but an unnecessary code for ordinary physicians.'[38] This also explains why various international instruments do not incorporate the Nuremberg Code, and those that do are in the context of the laws of war and humanitarian law. Physicians and members of the medical profession also resisted debates about the ethical aspects of the medical research arising from the trials. They would rather that it not be talked about because of their portrayal of the atrocities as an on/off event and as 'unique to one place or time'.[39] The moral issue was to keep the State from interfering in research.[40]

The atrocities in the East, for example – those that were perpetrated by Japanese forces against the Chinese – went without scrutiny. If the Japanese biomedical research elites and the 'freelance medical researchers' who also took the opportunity to exploit the war situation had been brought to trial, a restatement of the Nuremberg principles or a reinforcement of research ethics would have most definitely been the outcome. It would have kept alive the spirit of the Nuremberg Code in the East and perhaps had an effect on the way the Nuremberg Code was assimilated into the ethics of the medical research community. Taken together, both the Nuremberg Code and a Tokyo doctors' trial judgement could have laid the foundation of a truly international and global bioethical law. As the world entered the 1950s, the Cold War progressed and deepened, with priorities becoming very clear. The USSR was pitted against America, each of their ideologies claiming superiority and each wooing countries to join their respective blocs. The Non-Aligned Movement did not make much of an impact on the hard geopolitics of that era. International Law and human rights were tainted by ideological warfare. Developing countries used their resources in the UN and the international arena generally to fight for self-determination and independence from colonial powers. Their focus was on their economies and not on directly health and medical research. International bioethical law, based on the foundations of the Nuremberg Code, did not get the opportunity to develop.

During the Cold War period a focus on human experimentational medicine was found necessary and was used by the US military.[41] At the same time, the pubic health service was involved in human experimentation. In many instances the Nuremberg principles were not adhered to.[42] The Plutonium Files, published in 1999, exposed the case of 18 people – 17 American adults and one Australian child who were injected with plutonium without their knowledge or consent in the course of human experimentation by the US defence research establishment. The experiment was conducted in the mid 1940s by the nuclear weapons group to determine the level of plutonium that could be safely handled by humans.[43] Protecting national security was put first and the protection of human rights was ignored.[44]

Progress of science and technology and national security concerns dominated the scene and a strong international regime underlined by ethical and legal documents would have helped address instances of violations of ethical principles in a much more comprehensive and viable fashion. But the resistance of physicians and scientists to be regulated by members outside the profession combined with the interest of various countries in human experimentation as part of the Cold War strategy helped to keep the Nuremberg Code and its principles from developing further and taking root as binding international law principles.

1.5 The metamorphosis of human experimentation

Concerns about the equitable and just regulation of human genomic research are deepened in the context of the changing nature of human experimentation. Since the 1970s research in general has greatly increased, and research on humans has diversified into different types – pharmaceutical research that tests new drugs, the application of other research protocols on humans, and the use of human material for research, including human tissue, and more recently the use of genetic material.

At the time when the Nazi doctors conducted their experiments they used human beings as subjects. In the years following the Second World War, human experimentation has been transformed because of technology. It would now be unrecognisable to the Nazi doctors because experiments are carried out at the molecular level, with no visible link with the human being, unlike the early years of human experimentation, where the 'whole' person had to be present for experiments. Today human tissue has replaced human beings in human experimentation,[45] highlighting the change in practice of human experimentation, where less and less material is being required from the humans[46] who take part in scientific experiments. Thus the visibility of the issue is reduced just as the stakes involved in human experimentation are getting higher. The information that can be obtained from trace amounts of human material is quite large. The number of experiments involving human beings has also increased manifold. As science and technology has come to occupy an important place in the growth of a nation, the place of humans in scientific experimentation has become commonplace thereby invoking the need for more rules to respond to the changing situation on human experimentation.

A further development which poses a lot of challenges to the regulation of human genomic research is the progress in engineering and software technology which has led to the development of highly sophisticated machines which combine increased mechanical power with increased computing power to generate models of human tissue and molecules. This trend, whereby technologies such as genomics, robotics, informatics and nanotechnology (research in extremely small-sized particles) are coming together in order to improve diagnosis, research and treatment of diseases leading to enhanced human capacities, is referred to as the convergence of technologies.[47] Technosciences such as biotechnology are even more difficult to regulate. As new ideas and concepts emerge there is a call for 'new culture which reflects the changing relationships between government, industry and techno-scientific development.'[48] This new culture is characterised by its global networks and global ways of knowledge production,[49] and is therefore attracting the attention of international lawyers.

1.6 The Ethical, Legal and Social Issues (ELSI) Research Program

The ELSI Research Program funded by the US National Institutes of Health (NIH) is one of the largest ethics programmes in the world and was an effort to address wider implications of human genomic research.[50] Combined with the lack of international deliberation on medical research and development of technology[51] and the rapidly changing nature of human experimentation, the even more complex issues of equity and justice in medical research kept getting little or no attention. When the HGP was started in the early 1990s, this fact did not escape the visionary approach of James Watson, who personally ensured the setting aside of 3–5 per cent of the project's budget for identifying and addressing the 'ethical, legal and social issues (ELSI)' of such research. Hailed as the first of its kind to be incorporated into a science project, it set the stage for the international governance of the human genome. Watson took the view that the scientific advances of genomics should be accompanied by an understanding of the knowledge so that it would not give rise to fear among people. The ELSI Research Program started as a fresh project and welcomed proposals for issues that were regarded as problematic and challenging in the context of the human genome.

The ELSI Research Program had a wide remit, yet the focus of the themes that were identified were still not wide enough. The major themes are listed below:[52]

> **Fairness in the use of genetic information** by insurers, employers, courts, schools, adoption agencies, and the military, among others.
>
> **Privacy and confidentiality** of genetic information.
>
> **Psychological impact and stigmatization** due to an individual's genetic differences.
>
> **Reproductive issues** including adequate informed consent for complex and potentially controversial procedures, use of genetic information in reproductive decision making, and reproductive rights.
>
> **Clinical issues** including the education of doctors and other health service providers, patients, and the general public in genetic capabilities, scientific limitations, and social risks; and implementation of standards and quality-control measures in testing procedures.
>
> **Uncertainties** associated with gene tests for susceptibilities and complex conditions (e.g. heart disease) linked to multiple genes and gene-environment interactions.
>
> **Conceptual and philosophical implications** regarding human responsibility, free will versus genetic determinism, and concepts of health and disease.
>
> **Health and environmental issues** concerning genetically modified foods (GM) and microbes.
>
> **Commercialization of products** including property rights (patents, copyrights, and trade secrets) and accessibility of data and materials.

The ELSI Research Program did not tackle questions about the scientific project as a whole – the ethics of doing 'big science', the ethics of a 'fast track to knowledge' or the moral justification of resource allocation. There was a built-in understanding into the programme that all genomics knowledge is good. The understanding is built on Watson's logic, which is to 'Just do good, and [not] care if it doesn't seem good to others.'[53] Watson, however, strategically always encouraged public debate on science. The ELSI Research Program was criticised by many as a 'political pre-emptive strike intended to thwart criticism of science'.[54]

Robert Cook-Deegan has succinctly captured the central issue in genomics that is of great concern at the global level. He points to the extravagance of the claims that the HGP will help the diseased and the poor in his criticism of a claim (in the context of mental illness) that prevention of mental illness (by identifying and finding a cure for the genes that cause such illness) should be the focus, rather than researching into better means of caretaking.[55] While acknowledging the potential of knowledge from the HGP to alleviate human suffering, Cook-Deegan rebuts the idea that benefits will automatically accrue from the effort and that an understanding of genetics 'does not run a direct course from a gene to homelessness'.[56] There is a tendency to present long-term benefits tagged to short-term benefits, thus skewing the whole picture of benefits, which obfuscates the understanding of the public and the policy maker alike. In a report published in 2002, the World Health Organization, an intergovernmental organisation and a Specialised Agency of the UN, investigated the kind of benefits that might be expected from the study of the human genome and the identification of the functions of genes. The authors concluded that the impact of such research would only deliver benefits to world health if there was a particular focus on projects that addressed health issues that affected populations in countries with low economic growth and numerous health problems.[57, 58]

Despite the shortcomings of the ELSI Research Program and the narrow focus of the themes considered within it, if such programmes are rolled out on a global level they can become the thin end of the wedge; a powerful forum where the protection of human rights, particularly those of the disadvantaged and the unrepresented are granted primary importance. In the European Union, consideration of social, legal and ethical aspects of genomics (dubbed ELSA) was on the horizon in early genomic times due to the efforts of Germany. The presence of an ELSA programme alongside the scientific effort added weight to the whole proposal and the scientific project was approved. The ELSA programme was allocated 7 per cent of the HGP budget but still failed to explore issues of significance to global health.[59] The Japanese Ethics Programme, which began in 1991, was a small one, with an annual budget of 67,000 USD.[60] In Japan, corporate interests were predominant, followed by scientists' interests and then ethics. The Canadian ELSI programme set aside 7.5 per cent of the budget for ELSI issues.

1.7 Property rights in the human genome

Ownership and property rights in genetic material are more clear-cut arenas of contention in international law regarding genetic resources[61] as well as within European and municipal legal systems. The Convention on Biological Diversity[62] confers sovereign rights on plant biological resources.[63] The Agreement on Trade-Related Aspects of Intellectual Property Rights[64] (TRIPS) seeks to harmonise intellectual property rights (IPR) protection across various jurisdictions, and in Article 27 provides that 'patents shall be available for any inventions, whether products or processes, in all fields of technology, provided that they are new, involve an inventive step and are capable of industrial application'. According to subsection 2 of this Article, States Parties can exclude from patentability inventions that they consider are contrary to '*ordre public* or morality, [so as] to protect human, animal or plant life or health or to avoid serious prejudice to the environment, provided that such exclusion is not made merely because the exploitation is prohibited by their law'. Both plant genetic resources and human genetic materials have been subject to IPR and their protection or lack of it has given rise to disputes, with varying results as to the application of the fundamental principles of intellectual property (IP) law. The disputes have gone to the very heart of the purpose and objective of IP regimes and have essentially opened the notions of property and ownership to debate.

Many indigenous populations have unique genetic make-up, which makes their blood or other samples useful and important for medical research. An equitable regime for human genome research should be underpinned by a just and equitable system of ownership so that their rights are adequately protected. The question arises as to whether it is right to attach property rights to DNA sequences,[65] which is tantamount to granting rights on one's heritage. This question arose in the Moore case in the US,[66] where John Moore was being treated for a medical condition. It was held that although Moore was the supplier of the tissue specimen (from which genetic information was gathered) which was used to develop a product that was then the subject of a successful patent application, Moore himself could not claim any property rights in the sample because body parts and samples cannot be the subject of ownership. No individual can claim a direct property right over their genetic information. But it is possible to obtain rights that are conferred via the protection of IP, so 'ownership is not related to the human from whom the substance is taken; the property in the substance is created in the act of isolating the substance',[67] thus disenfranchising the individuals whose genetic information it is and benefiting patent holders, and other entities that could exploit the patent. While the fruit of labour theory applies to the isolation of the substance, it does not seem to apply to the formation of the substance which is in any case linked to a number of different factors and ultimately to the fact that human genome has been commonly shaped by humankind as a whole, and by various generations of them.

The Oncomouse case[68] and the Anant Chakrabarthy[69] case, which involved the patenting of whole organisms, raised fears about the patenting of life forms.[70] Here a genetically modified mouse and a microorganism were the subjects of patent protection. But in the absence of any other effective model of research, the general trend of granting exclusive property rights for a relatively limited period of time (in most cases 20 years) to products of highly expensive research such as genomic science, is the current model being pursued. Developing models for access to genetic research and the sharing of benefits from such research now occupies various academic research groups[71] and policy makers at intergovernmental organisations[72] like the World Intellectual Property Organization (WIPO).

The Hagahai are a group of people living in Papua New Guinea, consisting of just 260 members, who were isolated from the rest of world until recently. The US NIH filed a patent on the DNA from their blood samples as a means to create a vaccine against a human disease. A patent granted to the US Department of Health by the US patent office in 1995 on the genetic information of a Hagahai tribesman raised concerns about the ownership of information and rights of indigenous people.[73] The patent has since been relinquished. The Human Genome Diversity Project (HGDP) was launched by a group of scientists from Stanford University in 1991 to study variation in human groups across the world by tracing the genetic similarities and differences between them.[74] The proposal was to collect samples of DNA from various populations around the world and store them in regional DNA banks and also at a central database, which could then be used for research by scientists across the globe. It was not designed as a commercial project and was intended to provide geneticists with research material to further the genomics era. But the project lost momentum[75] because of criticisms arising from its focus on indigenous populations.[76] A similar project, known as the Genographic Project, involves many of the HGDP principal investigators and was initiated in April 2005, sponsored by the National Geographic Society, IBM and the Waitt Family Foundation. The aim of the project is to 'embark on a landmark five-year study that will assemble the world's largest collection of DNA samples to map how humankind populated the planet'. It proposes to map human migratory patterns over 150,000 years by collecting and analysing DNA blood samples from over 100,000 indigenous people making it the largest study of its kind in the field of anthropological genetics.[77] This too has been criticised.[78] In the light of the increasing importance of genomic information for a variety of different purposes ranging from addressing health issues to human migratory patterns, the protection of the interests of vulnerable groups, of which indigenous populations are one example, acquires importance.[79] Is it fair to attach property rights only to certain types of labour as opposed to others?

Resolution of the issues on ownership and property rights is a crucial first step in moving towards a more equitable regime for medical research in general and genomic research in particular. IPR protect moral interests, not just economic interests.[80] Moral issues have engaged nations since the Second World War in different ways, and have tended to be addressed via human rights law and the philosophy underpinning such law. A similar approach will be adopted in the chapters to come in addressing issues on property and IP in order to bring them closer to considerations of fundamental importance to human flourishing. The challenge that genomic research is presenting policy makers and thinkers with can be robustly met by exploring the depths of human rights philosophy to discover the approaches that can be adopted towards such diverse issues as health, property and IP.

The pace of progress in science is fast increasing, leaving the development of international rules lagging behind. Scientists are now studying the human proteome. The proteome is the sum total of all proteins that are coded for in the human genome. The value in decoding the human genome lies in uncovering the proteins that it can produce using the encoded information that it carries, in other words, it is the 'business end of biology'.[81] Nanotechnology is a new science which explores particles at a very minute scale and analyses particle behaviour. Nanotechnology in combination with biotechnology is expected to bring about a revolutionary change in many fields. The rapid progress of new technologies, the change in funding patterns, and the media attention to scientific developments have all transformed scientific research. The culture of scientific research, once one of open sharing of information, has now become more competitive and restrictive, bringing with it many ethical issues.

1.8 Human rights – central to governance of human genome

The governance of human genome research has been much debated by bioethicists operating within the national and the regional levels and some specific issues such as organ trade are the subject of international debate. How then can human rights law claim centrality in the regulation of this new field of research?[82] What are the connections between human rights and bioethics?

1.8.1 Bioethics

An early definition of bioethics is that it is a 'new discipline that combines biological knowledge with a knowledge of human value systems'.[83] Further elaborated, bioethics is 'the systematic study of the moral dimensions – including moral vision, decisions, conduct and policies – of the life sciences and health care, employing a variety of ethical methodologies in an interdisciplinary setting'.[84]

As Arthur Kaplan puts it, 'bioethics was born from the ashes of the Holocaust.'[85] International lawyers would state that human rights were born from the ashes of the Holocaust too. And both would be true. Bioethical debate and human rights theory are inextricably linked. The Nuremberg Code is both an instrument of bioethics and human rights law. Yet bioethics is alien to the international lawyer because, recast as a human rights issue, its nature would change. This is so primarily because of the nature of the human rights issues thrown up by bioethics and the new entities being introduced.

The classic text on bioethics was written by Beauchamp and Childress[86] in 1979 and is based on American experiences. The Tuskegee Syphilis Study in America and the thalidomide drug case in Europe[87] led to widespread discussion on ethics and medicine. The Tuskegee research project brought to light the callousness of some members of the medical profession who, in conjunction with the government, were involved in 'bad science'. The research project, which spanned from 1932 to 1972, was government-run, in a small town in America, using human subjects (African Americans) who carried syphilis to study the effect of the disease on them. The study subjects were not told the purpose of the study, their consent was not obtained and they were not treated, resulting in the death of many participants. The government finally ended the experiment in 1972 because of media publicity and the remaining subjects were treated for the disease, thus ending the longest ever human experimentation in medical history.[88]

The Tuskegee Study led to the setting up of the National Commission for the Protection of Human Subjects of Biomedical and Behavioral Research with a specific mandate to deal with requirements for informed consent, among other tasks. Its subsequent report, *Ethical Principles and Guidelines for the Protection of Human Subjects of Research*, known as *The Belmont Report*, was published in 1979 and alluded to the Nuremberg guidelines on human subject research and went on to lay down two categories of guidelines. The authors enumerated three basic principles, namely respect for persons, beneficence and justice, and three applications of these principles, which listed informed consent as the first application of the general principles.[89] These principles form the basis of the development of bioethical principles in the US, of which Beauchamp and Childress were advocates. Their four-principle approach to[90] bioethical dilemmas became the founding principles in bioethics.[91] They are:

1 Autonomy
2 Non-maleficence
3 Beneficence
4 Justice.

The principle of beneficence dictates that health care providers are obligated to help people in need; that of non-maleficence instructs physicians to do no harm; that of autonomy refers to the right of a person to make his or her own decisions; and the principle of justice requires the fair treatment of all persons.[92] One commentator has observed that these 'principles are derived from common morality as opposed to reason or natural law, but they still have deep justification in philosophy.'[93] Another is of the view that they are 'mid-level principles mediating between high-level moral theory and low-level common morality.'[94] Writing about the four-principle approach, one of the authors of the text observes that 'very general moral principles are shared by all moral persons, but that specified moral frameworks (those that moral agents likely bring to and develop in case analysis) are works in progress that can legitimately vary from person to person and from culture to culture.'[95] He notes that 'specified moralities include the many moral norms, aspirations, ideals, attitudes, and sensitivities that spring from cultural traditions, religious traditions, professional practice, institutional codes of ethics, and the like.'[96]

As bioethical debate has spread and developed across various parts of the globe, it has widened in scope and includes a wide range of perspectives.[97] In essence this field of enquiry is interdisciplinary and rapidly evolving, embedding features of international human rights law, and could probably be classified as a branch of human rights law. However, although human rights principles and law is central to the regulation of research on the human genome, in practice the nature of international human rights law itself needs to be called into question in order to address the issue of how it can contribute to the governance of the human genome. For example, human rights include the right to health, and the factors that are responsible for the wellbeing of peoples – bioethics is also concerned with the same issues. But to what extent is the right to health well-established in international human rights law? Furthermore, human rights law and bioethics differ 'in scope, sources [and] legal nature'.[98]

The notion of human rights leans heavily towards civil and political rights. International human rights law has had more success in protecting the right to a fair trial, the right to vote, the right to freedom of speech and freedom of movement and the right to have a nationality than it has had with protecting the right to food, the right to work, the right to social security, the right to health and other economic, social and cultural rights (ESCR). On the one hand, civil and political rights stand out as rights that are conferred on individuals by international instruments; on the other hand, economic, social and cultural rights come across as 'inchoate rights, which are the product of reasoned moral analysis.'[99] The Article 2 provision of 'progressive development' and a lack of effective monitoring of ESCR have led to a second-class status for ESCR. Even the right not to be subjected to torture through the use of scientific experimentation is provided in the context of first generation human rights, although the effect of scientific experimentation is first and foremost on the health of the individual.

Notes

1 E. Pennisi, 'Finally the Book of Life and the Instructions for Navigating It', *Science* 288 (5475), 2000, 2304–7.

2 Dr Michael Morgan, Chief Executive of the Wellcome Trust Genome Campus, *The Future is Genomic*, Wellcome Trust Sanger Institute, available online at http://www.sanger.ac.uk/HGP/draft2000/future.shtml (accessed 8 April 2006); Stephen S. Hall, 'James Watson and the Search for Biology's "Holy Grail"', *Smithsonian* 20 (11), 1990, 40–9.

3 'We've now got to the point in human history where for the first time we are going to hold in our hands the set of instructions to make a human being.' John Sulston, Director of Sanger Centre, Cambridge, UK, who was the team leader of the Cambridge sequencing group, interviewed by the BBC, 26 June 2000. Available online at http://news.bbc.co.uk/2/hi/science/nature/807126.stm (accessed 18 August 2008).

4 A. Mauron, 'Is the genome the secular equivalent of the soul?', *Science* 291 (5505), 2001, 831–2 at 831.

5 G. J. Annas, 'Governing Biotechnology', *Global Agenda Magazine*, available online at http://www.globalagendamagazine.com/2006/Annas.asp (accessed 6 April 2006).

6 J. D. Watson, F. H. C. Crick, 'A Structure for Deoxyribose Nucleic Acid', *Nature*, 1953, 171–7.

7 IBM Archives, available online at http://www-03.ibm.com/ibm/history/ (accessed 11 April 2006).

8 For a brief history see Y. Segal, *The Israel Academy of Sciences and Humanities*, available online at http://www.daat.ac.il/daat/kitveyet/assia_english/segal.htm (accessed 7 May 2006).

9 US Department of Energy Office of Science, Office of Biological and Environmental Research, Human Genome Program, 'About the Human Genome Project', 2008, available online at http://www.ornl.gov/sci/techresources/Human_Genome/project/about.shtml

10 R. Cook-Deegan, *The Gene Wars: Science, Politics and the Human Genome*, New York: W. W. Norton & Company, 1994, p. 215.

11 R. R. Carp in G. Kolata, 'Biology's Big Project Turns into Challenge for Computer Experts, *The New York Times*, 11 June 1996, p. C12, cited in J. Rifkin, *The Biotech*

Century: Harnessing the Gene and Remaking the World, London: Victor Gollancz, 1998, p. 191.

12 R. Cook-Deegan, *The Gene Wars: Science, Politics and the Human Genome*, New York: W. W. Norton & Company, 1994.

13 Speaking at a Third Committee Meeting at the UN General Assembly, the representative of India stated 'Unfortunately, Social Darwinism (is) back with the advent of the mapping of the human genome'. He also said 'racial distinctions continued to be driven by the false assumption that human beings could be divided into sub-species according to their genetic characteristics' (56th General Assembly, Third Committee, 1 February 2002, 59th Meeting. Also see the speech by Mary Robinson, UN High Commissioner for Human Rights, given at the European Conference Against Racism, Strasbourg, 11–13 October 2000 and the report of States Parties to the Committee on the Elimination of Racial Discrimination (a treaty body established under the Convention on the Elimination of All Forms of Racial Discrimination), CERD/C/BGR/19, 30 January 2008.

14 G. Gárdián, L. Vécsei, 'Huntington's disease: pathomechanism and therapeutic perspectives', *Journal of Neural Transmission* 111 (10, 11), 2004, 1485–94.

15 R. A. King, *et al.* (eds), *The Genetic Bases of Common Diseases*, Oxford: Oxford University Press, 2002.

16 James Watson, in particular, was a towering personality, under whose leadership the National Institutes of Health (NIH) became a world leader in DNA sequencing and Genomics. For more information see J. D. Watson, *The Double Helix*, London: Weidenfeld & Nicolson, 1968 (with a new introduction, 1997) and J. D. Watson, *The Molecular Biology of the Gene*, 1st edn, New York: W. A. Benjamin, 1965.

17 V. A. McKusick, 'The Human Genome Organisation: History, Purposes and Membership', *Genomics* 5 (2), 1989, 385–7, available online at http://www.hugo-international.org/abt_history.php (accessed 26 July 2008).

18 K. M. Nolan-Stevaux, 'Open Source Biology: A Means to Address the Access and Research Gaps?', *Santa Clara Computer and High Technology Law Journal* 23, 2007, 271–316 at 293.

19 Cook-Deegan, op. cit.

20 Fortun, M. A., 'Celera Genomics: The Race for the Human Genome Sequence' in *Encyclopedia of Life Sciences*, John Wiley & Sons, Inc., 2006. Available online at http://mrw.interscience.wiley.com/emrw/9780470015902/els/article/a0005182/current/abstract (accessed 12 March 2008); J. Shreeve, *The Genome War: How Craig Venter Tried to Capture the Code of Life and Save the World*, London: Knopf (Random House), 2004.

21 'Scientists crack human code', BBC, 26 June 2000, available online at http://news.bbc.co.uk/1/hi/sci/tech/805803.stm (accessed 6 June 2007).

22 See PubMed for a list of journals related to genetics and genomics.

23 C. R. Scriver, 'Presidential Address', *American Journal of Human Genetics* 401, 1987, 99–211.

24 H. T. King, Jr, Nuremberg and Sovereignty, *Case Western Reserve Journal of International Law* 28, 1996, 135–40.

25 Q. Wright, 'The Law of the Nuremberg Trial', *The American Journal of International Law*, 41 (1), 1947, 38–72 at 71.

26 P. J. Weindling, *Nazi Medicine and the Nuremberg Trials: From Medical War Crimes to Informed Consent*, London: Palgrave Macmillan, 2004.

27 B. McFarland-Icke, 'Nazi Medicine and the Nuremberg Trials', *Social History of Medicine* 18 (3), 2005, 523–4.

28 P. J. Weindling, *Nazi Medicine and the Nuremberg Trials: From Medical War Crimes to Informed Consent*, London: Palgrave Macmillan, 2004, pp. 257–93.

29 'Permissible Medical Experiments' in *Trials of War Criminals before the Nuremberg Military Tribunals under Control Council Law No. 10. Nuremberg, October*

1946–April 1949, Washington, DC: US GPO, 1949–1953; *Directives for Human Experimentation: Nuremberg Code*, US National Institutes of Health (NIH), available online at http://www.nihtraining.com/ohsrsite/guidelines/nuremberg.html (accessed 5 April 2006).

30 G. Annas, L. H. Glantz, B. F. Katz (eds), *Informed Consent to Human Experimentation: The Subject's Dilemma*, Cambridge, Massachusetts: Ballinger, 1977, p. 21.

31 See *United States v Brandt, United States v Flich* in *Trials of War Criminals before the Nuremberg Military Tribunals under Control Council Law No. 10. Nuremberg, October 1946–April 1949*, Washington, DC: US GPO, 1949–1953.

32 Ibid.

33 Ibid., p. 183.

34 In the US the first codes on experimentation came after the Nuremberg trials; see V. Orlowski, 'Promising Protection Through Internationally Derived Duties', *Cornell International Law Journal* 36, 2003, 381–413 at 389.

35 J. Huxley, *UNESCO: It's Purpose and its Philosophy*, Washington DC: Public Affairs Press, 1947, p. 14.

36 P. J. Weindling, *Nazi Medicine and the Nuremberg Trials: From Medical War Crimes to Informed Consent*, London: Palgrave Macmillan, 2004, p. 320.

37 For evidence to the contrary, see H.M. Hanauske-Abel, 'Not a slippery slope or sudden subversion: German medicine and National Socialism in 1933', *British Medical Journal* 313, 1996, 1453–63. In 1986 Dr H. Hanauske-Abel published an article in *The Lancet* on the role of German doctors under National Socialism. Within two weeks his sub-licence to practise emergency medicine was withdrawn by the Kassenarztliche Vereinigung (which is organisationally closely related to the Chamber of Physicians) and was restored to him only after a decision by the German Supreme Court. In 1987 Dr Hanauske-Abel brought a lawsuit against the Deutsches Arzteblatt (which has the same relation to the German Chamber of Doctors as the *British Medical Journal* has to the British Medical Association). He claimed that an interview with the president of the chamber on the topic of *The Lancet* article was defamatory. The courts found in favour of the defence.

38 J. D. Moreno, 'Reassessing the Influence of the Nuremberg Code on American Medical Ethics, *Journal of Contemporary Health Law & Policy* 13, 1997, 347–60 at 350.

39 Leaning J., 'War crimes and medical science', *British Medical Journal* 1996: 313 (7070), 1413–15.

40 P. J. Weindling, 'Human guinea pigs and the ethics of experimentation: the *BMJ*'s correspondent at the Nuremberg medical trial', *British Medical Journal* 313 (7070), 1996, 1467–70.

41 *Human Radiation Experiments Associated with the U.S. Department of Energy and Its Predecessors*, US Department of Energy, Assistant Secretary for Environment, Safety and Health, Washington, DC 20585, July 1995.

42 J. D. Moreno, 'Reassessing the Influence of the Nuremberg Code on American Medical Ethics, *Journal of Contemporary Health Law & Policy* 13, 1997, 347–60 at 351.

43 E. Welsome, *The Plutonium Files: America's Secret Medical Experiments in the Cold War*, The Dial Press, 1999.

44 G. J. Annas, 'Mengele's Birthmark: The Nuremberg Code in United States Courts', *Journal of Contemporary Health Law & Policy* 7, 1991, 17–45.

45 It has to be noted that a reductionist suggestion is not being made here that human tissue is equivalent to human beings; nor is there an argument to support the view that DNA, which is a special type of human tissue, is merely a chemical substance that can be treated as such.

46 For humans as wholes and parts see E. Giarelli, 'Safeguarding being: a bioethical principle for genetic nursing care', *Nursing Ethics* 10 (3), 2003, 255–68.

47 M. C. Roco, W. S. Bainbridge (eds), *Converging Technologies for Improving Human Performance: nanotechnology, biotechnology, information technology and cognitive*

science, Arlington, Virginia: National Science Foundation, 2002.

48 P. Glasner, P. Atkinson, 'The Global Genome: Biotechnology, Politics and Culture', *Body and Society* 12 (3), 2006, 123–31.

49 S. Jasanoff (ed.), *States of Knowledge: The Co-production of Science and Social Order*, London: Routledge, 2004.

50 'Ethical, Legal and Social Issues', The Department of Energy, Government of the United States of America, available online at http://www.ornl.gov/sci/techresources/Human_Genome/elsi/elsi.shtml (accessed 12 June 2007). The most extensive ELSI programme was the one funded by the US NIH and the Department of Energy. See A. Ari Patrinos, D. W. Drell, 'Introducing the Human Genome Project: Its Relevance, Triumphs, and Challenges', *Judges' Journal* 36 (3), 1997, 5–10.

51 Barring the few UNESCO resolutions from the 1960s onwards which make a general statement on progress in science and technology.

52 'Ethical, Legal and Social Issues', Human Genome Project Information, available online at http://www.ornl.gov/sci/techresources/Human_Genome/elsi/elsi.shtml (accessed 12 April 2006).

53 Cook-Deegan op. cit., p. 166.

54 Ibid., p. 248.

53 Ibid., p. 246. As D. Koshland has written, 'The costs of mental illness, the difficult civil liberties problems they cause, the pain to the individual, all cry out for an early solution that involves prevention, not caretaking. To continue the current warehousing or neglect of these people, many of whom are in the ranks of the homeless, is the equivalent of providing iron lungs to polio victims at the expense of working on a vaccine' (D. Koshland, 'Editorial: The Rational Approach to the Irrational', *Science*, 250, 1990, 189).

54 'Ethical, Legal and Social Issues', Human Genome Project Information, available online at http://www.ornl.gov/sci/techresources/Human_Genome/elsi/elsi.shtml (accessed 12 April 2006), p. 247.

55 Advisory Committee on Health Research, *Genomics and World Health*, World Health Organization, Geneva, 2002.

56 Also see R. Cook-Deegan, *et al.*, *World Survey of Funding for Genomic Research*, Stanford-in-Washington Program, Washington DC, 2000.

57 See the Council of European Communities' decision adopting a specific research and technological development programme in the field of health: EU Doc. 90/395/EEC, 1991.

58 R. Cook-Deegan, *The Gene Wars: Science, Politics and the Human Genome*, New York: W. W. Norton & Company, 1994, p. 226.

59 See the statements of members of the Sub-Commission on the Promotion and Protection of Human Rights in the Press Release, 54th Session, 15 August 2002, available online at http://www.unhchr.ch/huricane/huricane.nsf/0/31071122AE996F1BC1256C17002739D2?opendocument (accessed 9 August 2008).

60 1760 *United Nations Treaty Series* 79, reproduced in *International Legal Materials* 31, 1992, 818.

61 Article 3, Convention on Biological Diversity: 'States have, in accordance with the Charter of the United Nations and the principles of international law, the sovereign right to exploit their own resources pursuant to their own environmental policies, and the responsibility to ensure that activities within their jurisdiction or control do not cause damage to the environment of other States or of areas beyond the limits of national jurisdiction.'

62 15 April 1994, Marrakesh Agreement Establishing the World Trade Organization, Annex 1C, The Legal Texts: The Results of the Uruguay Round of Multilateral Trade Negotiations 320, 1999, 1869 *United Nations Treaty Series* 299, reproduced in *International Legal Materials* 33, 1994, 1197.

63 R. Hardcastle, *Law and the Human Body: Property Rights, Ownership and Control*, Oxford: Hart Publishing, p. 145.

64 *Moore v. Regents of University of California*, 1990, 793 P. 2d 479 and 1991, 111 S Ct 1388.
65 D. Townend, 'Who owns genetic information?', in J. Sandor (ed.) *Society and Genetic Information: Codes and Laws in the Genetic Era*, Budapest: CPS-Central European University Press, 2003, p. 132.
66 M. Swenarchuk, *The Harvard Mouse And All That: Life Patents In Canada*, CELA Publication #454, Toronto: Canadian Environmental Law Association, 2003.
67 *Diamond v. Chakrabarty*, 447 US 503, 1980.
68 J. Wild, 'The Future for Patents on Life', 2002, available online at http://scientific.thomson.com/free/ipmatters/patlife/8180006/ (accessed 13 October 2006).
69 D. Schroeder, *et al.*, 'Sharing the benefits of genetic research', *British Medical Journal* 331, 2005, 1351–2.
70 The General Assembly highlights the main concerns in a resolution adopted in 1994 in which member States 'emphasised the fact that many advances in scientific knowledge and technology in health education, housing and other social spheres should be readily available to the populations as the heritage of humanity, with a view to sustainable development, taking into account the need to protect intellectual property rights' (UN Doc. A/RES/48/140,1994).
71 O. Dyer, 'US awards patent for tribesman's DNA', *British Medical Journal*, 311, 1995, 1452.
72 See the HDGP page of the Stanford University website at www.stanford.edu/group/morrinst/hgdp.html (accessed 12 April 2006). See also A. M'Charek, *The Human Genome Diversity Project: An Ethnography of Scientific Practice*, Cambridge: Cambridge University Press, 2005.
73 See L. Cavalli-Sforza, 'The Human Genome Diversity Project: past, present and future', *Nature Reviews Genetics* 6 (4), 2005, 333–40.
74 J. Marks, 'Human Genome Diversity Project: Impact on Indigenous Communities', in D. Cooper (ed.), *Encyclopedia of the Human Genome* (5 vols), London: Macmillan, available online at http://personal.uncc.edu/jmarks/pubs/main.html (accessed 12 April 2006).
75 The Genographic Project, available online at https://www3.nationalgeographic.com/genographic/faqs_about.html#Q1 (accessed 12 April 2006).
76 *Consolidated Indigenous Shadow Report – Examination of the United States' 4th, 5th and 6th Periodic Reports of April 2007*, submitted to the Committee on the Elimination of Racial Discrimination, p. 60.
77 See the statement of Chris Chapman of Minority Rights Group International at the 55th Session of the Sub-Commission on the Promotion and Protection of Human Rights, 11 August 2003, available online at http://www.unhchr.ch/huricane/huricane.nsf/0/670B54E219DC300DC1256D7F004FDC49?opendocument accessed (accessed 11 August 2008).
78 H. Macqueen, *et al.*, *Contemporary Intellectual Property: Law and Policy*, Oxford: Oxford University Press, p. 8.
79 'Editorial: Proteomics' new order', *Nature* 437, 2005, 169–70.
80 E. Petersmann, 'Biotechnology, Human Rights and International Economic Law' in Francioni F. (ed), *Biotechnologies and International Human Rights*, Oxford: Hart Publishing, 2007, 229–73, p. 232.
81 A. R. Jonsen, *The Birth of Bioethics*, New York: Oxford University Press, 1998, p. 27.
82 W. T. Reich, 'Introduction', in Reich (ed.), *The Encyclopedia of Bioethics*, rev. edn, New York: Simon Schuster Macmillan, 1995, p. xxi.
83 G. J. Annas, 'American Bioethics and Human Rights: The End of All Our Exploring', *Journal of Law, Medicine & Ethics* 32, 2004, 658–62 at 659.
84 T. L. Beauchamp, J. F. Childress, *Principles of Biomedical Ethics*, New York: Oxford University Press, 1st edn, 1979 (5th edn, 2001).

85 The thalidomide drug, invented by a West German company, was released in 1957 under various names in European countries. Its release in other countries including Japan and Australia soon followed. It was taken by women during pregnancy and resulted in the birth of children with deformities. Legal action was taken in courts across the world leading to compensation for the victims of the drug.

86 J. H. Jones, *Bad Blood: The Tuskegee Syphilis Experiment*, Free Press, rev. edn, 1993.

87 The National Commission for the Protection of Human Subjects of Biomedical and Behavioural Research, *The Belmont Report: Ethical Principles And Guidelines For The Protection Of Human Subjects Of Research*, Department of Health, Education, and Welfare, United States of America, 18 April 1979, available online at http://ohsr.od.nih.gov/guidelines/belmont.html#top (accessed 11 April 2006).

88 See *International Guidelines for Ethical Review of Epidemiological Studies*, Council for International Organizations of Medical Sciences (CIOMS), 1991, which embodies the four principles and is available online at http://www.cioms.ch/frame_1991_texts_of_guidelines.htm (accessed 11 January 2006).

89 See also R. Gillon, 'Medical ethics: four principles plus attention to scope', *British Medical Journal* 309, 1994, 184 for an analysis supporting these four principles as a culturally neutral approach to thinking about ethical issues in health care and R. Gillon, 'Ethics needs principles – four can encompass the rest – and respect for autonomy should be "first among equals"', *Journal of Medical Ethics* 29, 2003, 307–12 for a discussion on hierarchy in the four principles.

90 L. Birnbaum, Medical Ethics in Exercise Physiology, *Journal of Professional Exercise Physiology* 10 (3), 2005.

91 R. J. D. Macer, *Bioethics is Love of Life: An Alternative Textbook*, Christchurch, New Zealand: Eubios Ethics Institute, 1998. eBook available at http://www2.unescobkk.org/eubios/BLL.htm

92 S. Holm, 'Book Review: Principles of Biomedical Ethics, 5th edition', *Journal of Medical Ethics 28*, 2002, 332.

93 T. L. Beauchamp, Methods and Principles in Biomedical Ethics, *Journal of Medical Ethics* 29, 2003, 269–74.

94 Ibid.

95 Macer puts forward the understanding that bioethics is love of life in *Bioethics is Love of Life: An Alternative Textbook*, Christchurch, New Zealand: Eubios Ethics Institute, 1998. eBook available at http://www2.unescobkk.org/eubios/BLL.htm

96 S. P. Marks in S. Post (ed), *Encyclopedia of Bioethics*, 3rd edn (5 vols), Macmillan Reference, 2004, vol. 2, pp. 1221–7.

97 I. Kennedy, 'Patients, doctors and human rights', in R. Blackburn, J. Taylor (eds), *Human Rights for the 1990s*, London: Mansell Publishing Limited, 1991.

2 International organisations and the human genome

2.1 Introduction

UNESCO, the Human Rights Council (HRC), the European Union, the World Health Organization (WHO) and the General Assembly have been engaged in developing international law to regulate biotechnology.

The phrase 'protect and preserve'[1] has been employed to describe the international community's engagement with the human genome, but in reality it is still a matter of debate as to what it is that the international community is trying to do with the human genome. It would be impossible to 'preserve' the genome, since it is a changing entity, one that is constantly evolving and adapting. And 'protection' is a malleable term when applied to the genome. First there is the question of what it is that is being protected. The second question is about the nature of the threat and the perpetrators. Third, is it the genome that needs protection or is it the human beings (in and for whom the genome exists) that are the ones who will be affected by any threats to the human genome? The current state of the debate on these issues is that we are still trying to come to grips with these fundamental issues. In this context, the spirit of the 1997 Universal Declaration on the Human Genome and Human Rights – that the human genome is the common heritage of mankind (CHM) – makes a simple yet far reaching statement that all these questions should be debated and decided upon, in the interest of humanity.[2]

Of the Specialised Agencies of the UN, UNESCO has been involved for the longest period of time in the promotion of scientific research on the human genome and the regulation of such research. Its initial involvement was in bringing about international cooperation in genomic research and to facilitate the training of scientists from developing countries who would otherwise not be able to acquire access to cutting-edge genomic technology. In 1989[3] and 1990 two international conferences were organised by UNESCO under the active leadership of Federico Major, the Director-General of UNESCO, who was himself a biochemist. The 2nd International Conference on Genomics was held in Valencia in November 1990 and concentrated on ethical issues. Aware of other programmes such as the ELSI Research Program (researching the ethical, legal and social issues of the HGP) in the US, UNESCO sought to cultivate a dialogue on the ethics and regulation of genomic research at the international level. The encouragement of capacity-building in Third World countries and the establishment of North-South genomics networks and regional networking through conferences and meetings also underscore the importance of populations in the Third World for human genomic research.

In 1990, the Council for International Organizations of Medical Sciences (CIOMS) adopted the Declaration of Inuyama on Human Genome Mapping, Genetic Screening and Gene Therapy.[4] The international conference at which the Declaration was adopted was held under the auspices of the Science Council of Japan, and cosponsored by WHO and UNESCO. It was the fifth in a series entitled Health Policy, Ethics and Human Values: An International Dialogue, begun in Athens in 1984. The participants, numbering 102, came from 24 countries, representing all continents. The organisations declared that there were no inherent ethical problems in the efforts to map the human genome. It urged States to ensure 'the setting and observance of ethical standards through continuous multidisciplinary and transcultural dialogue'. The possibility that human genome mapping and the subsequent benefits in terms of screening and therapy could pass large populations in developing countries by was acknowledged in the Declaration of Inuyama.[5]

2.2 UNESCO and bioethics

UNESCO has initiated and sustained an international dialogue on the ethics of science and technology since its inception in 1945. In recent years, the Division of Social and Human Sciences has been actively engaged in encouraging debate on new and emerging technologies in various fields, such as biotechnology, nanotechnology, space exploration, information technology, etc.[6] It has sought to provide an international focus on the issues concerned and has often dealt with issues that have been neglected in a national setting, where only those issues that affect the majority of that population gets most attention. In 1993, the Secretary-General established the International Bioethics Committee (IBC) to 'follow progress in the life sciences and its applications in order to ensure respect for human dignity and freedom'.[7] Since then, the IBC has been institutionalised into UNESCO by means of a statute.[8] It has adopted numerous standard-setting instruments in the field of biotechnology.

The Universal Declaration on Human Rights and the Human Genome was adopted by the General Conference of UNESCO in 1997, after much deliberation and discussion at the IBC. In 1999 an Intergovernmental Committee on Bioethics was established to examine the advice and recommendations of the IBC.[9] Together they drafted the 2003 International Declaration on Human Genetic Data[10] and the 2005 Universal Declaration on Bioethics and Human Rights.[11]In 1998 UNESCO established the World Commission on the Ethics of Science and Technology (COMEST) 'to formulate ethical principles that could provide decision-makers with criteria other than purely economic'. COMEST is an advisory body composed of independent experts.[12] At the initiative of the Director-General of UNESCO, the UN Inter-Agency Committee on Bioethics was established in 2003, with UNESCO acting as its permanent secretariat. It was established to reduce duplication by improving the coordination of the activities carried out by the following organisations that are working in the field of bioethics such as Office of the High Commissioner for Human Rights, the UN Food and Agriculture Organization (FAO), the International Labour Organization (ILO), UNESCO, WHO, the World Intellectual Property Organization (WIPO), the Arab League Educational, Scientific and Cultural Organization (ALESCO), the African Union, the Council of Europe, the European Commission, the Organisation for Economic Co-operation and Development (OECD), the World Trade Organization (WTO) and the International Centre for Genetic Engineering and Biotechnology (ICGEB).[13] UNESCO has published numerous books, studies and reports in ethics. In 2003 the Islamic Republic of Iran instituted the Avicenna Prize for Ethics in Science, now jointly sponsored by UNESCO and Iran for 'Ethics in Science'.[14]

2.2.1 The International Bioethics Committee

The IBC is a committee of 36 experts appointed by the Director-General of UNESCO on 15 September 1993 for a period of four years. The IBC was established following a General Assembly resolution dated 15 November 1993, which invited the Director-General 'to continue in 1994–1995 the preparation of an international instrument for the protection of the human genome'. The members of the IBC are chosen to fulfill the criteria on cultural diversity and balanced geographical representation through nominations from States of qualified specialists in the life sciences and in the social and human sciences, including law, human rights, philosophy, education and communication. They are experts in their fields and are of important national and international standing, functioning as judges, doctors, professors, directors of institutes, ambassadors and politicians and are drawn from countries across the globe. The composition of the IBC is adjusted to include equal representation to developing and developed countries and different cultures. They meet at least once a year and 'produce advice and recommendations on specific issues that are adopted by consensus and are widely disseminated and submitted to the Director-General for transmission to the Member States, the Executive Board and the General Conference'.[15] The Legal Commission of the IBC is composed mainly of jurists and lawyers from different backgrounds. Their main function, between 1993 and 1997, was to draft a legal instrument to govern the human genome, which was then subject to discussions at the IBC. The instrument was then sent to interested bodies around the world during a consultation period and was debated by the group of intergovernmental experts before it was finalised. The IBC can make recommendations within the areas of its competence and report to the Director-General who then decides on whether such recommendations are to be taken further to the plenary membership of UNESCO. These statements are of limited legislative authority, but have influence on behaviour in various ways.[16]

2.2.2 Drafting an instrument to protect the human genome

New developments in science and technology have simultaneously prompted awe and fear of the unknown. The high profile given to the Human Genome Project (HGP) heightened collective concerns about eugenics and other forms of exploitation, not to mention the horror of some, at others trying to 'play God'. Various professional organisations, international agencies and associations began to consider the impact of the HGP.

The World Medical Association (WMA), which represents the interests of physicians from all around the world, called the HGP a new diagnostic and knowledge-gathering tool. In a resolution adopted at the 44th World Medical Assembly in 1992, the World Medical Association (WMA) alluded to the power of this new tool but recommended that the use of the 'same parameters that guide us whenever we examine a new diagnostic or therapeutic method' should be used to evaluate this new technology.[17] It recommended the exchange of 'international information and transfer of technology and knowledge between all countries' and cautioned against increased social inequality resulting from inaccessible genetic services arising out of expensive techniques derived from the HGP.

The process of governance began in 1993 when the president of the IBC submitted a report recounting the involvement of UNESCO in the HGP and the subsequent developments in bioethics within UNESCO.

The President of the IBC, Dr Noelle Lenoir, recommended that the preparatory phase for an international standard-setting instrument on the human genome be commenced, with a call for an international conference of governmental experts to draft such an instrument for the protection of the human genome.[18] A resolution was passed by the General Conference of UNESCO in November 1993 recommending that work begin on an international standard-setting instrument in relation to the human genome. The resolution sought to apply human rights law to bioethics issues. Addressing the main challenges in drafting such an instrument, the Preamble emphasised 'the need to ensure the participation of all in the advances of the biomedical and life sciences and in the resultant benefits, with due regard for the freedom, dignity and identity of the human person'. Several resolutions of this nature were passed in the following sessions of the General Conference, urging and encouraging the IBC to progress with its work programme to draft an instrument to set standards for matters relating to the human genome.[19]

2.2.3 IBC Sessions: from 1993 to 1997

Human rights law was seen as the primary mechanism through which science and technological progress could be regulated. Progress in science and technology was welcomed with the general feeling that a new generation of human rights was in the making[20] and caution was needed in considering such progress as an unalloyed good. Freedom of research was held to be a fundamental human right, and the right to health was identified as being very important.[21] The human rights model of governance whereby the Universal Declaration was followed by the covenants, i.e. from guidelines to binding instrument, was favoured by most members. The international dimension of the HGP and the international nature of bioethics were acknowledged.[22]

Discussions at the IBC were underpinned by the notion that cultural diversity exists and that mutual understanding is required to promote a minimum consensus.[23] Specific applications were discussed. Among them were therapeutic applications stemming from genetic engineering; genetic screening and testing, access to genetic services and treatment, question of property in genetic information, confidentiality and genetic identity.[24] Diplomats, UN agencies such as the WHO, the Council of Europe, the Commission of the European Community, the OECD, the Pan-American Health Organization (PAHO) and the International Cancer Research Centre were present at the IBC. International non-governmental organisations such as International Social Science Council, CIOMS and the Human Genome Organisation (HUGO) had input into the IBC proceedings, either through the presence of a representative at the meetings or through written statements.

During the second meeting of the IBC in 1994, detailed presentations on numerous bioethics issues were given by the members of IBC.[25] Topics covered included population genetics (the new project known as the Human Genome Diversity Project (HGDP) was introduced and its objectives explained) and the teaching of bioethics. Therapeutic applications from genetic engineering and genetic screening and testing issues were further elaborated. Two documents were presented at this session entitled the 'Report on the draft for an international instrument for the protection of the human genome' and the 'First outline of a declaration on the protection of the human genome' by the Legal Commission. Attention was drawn to issues such as the granting of legal personality to the human genome[26] and the introduction of the CHM as a legal principle relating to the human genome.[27] Referring to the CHM concept, one IBC member and the then President of the International Court of Justice (ICJ) stated:

> What it means is that, looking beyond individuals, social groups or the State, we must safeguard the higher or transcendental interest of humanity as a whole and share with one another the resources of this and any other planet, in other words all life's blessings, in a spirit of human solidarity extending beyond borders.[28]

The WMA and a number of representatives of the national and international press attended the 1994 proceedings.

In the Third Session (1995), the draft for the instrument for the protection of the human genome, prepared by the Legal Commission, was debated at the plenary sessions of the IBC. Issues that figured prominently on the draft were the CHM, requirement for free and informed consent in research, liability issues, and prohibitions on various biomedical procedures considered abhorrent to human dignity.[29] A session on the role of parliaments for regulating new genetics issues, or new developments in science and technology, discussed the relative merits of national and international law.[30] While commenting on the CHM, the Director-General of

UNESCO noted that international law made use of this concept in the context of culture and human rights.[31] He was referring to the Convention Concerning the Protection of the World Cultural and Natural Heritage, which classifies cultural and natural heritage sites as the CHM. He further stated that the concept should be 'maintained in the declaration in view of the need to guarantee respect for human dignity and human rights and the need for a balance between the protection of individual rights and the common interest of humanity.'[32] In the general debate that ensued, the private and public aspects of law were enunciated in the context of the CHM. The issue of intellectual property (IP) law was one of the private law issues discussed in this context and the President of the IBC, Dr Lenoir, stated that 'the "common heritage of humanity" was not incompatible with the prospect of the patentability of the human genome. The common heritage concept simply aimed to assert the principle of freedom of access to the results of genome research.'

Dr Lenoir was of the opinion that UNESCO could not go any further in the Universal Declaration with regard to the question of patentability in view of the many interests at stake. She proposed, on the other hand, that in-depth examination of the matter be continued in the context of the IBC's work on ethics and genetic data banks. Freedom, responsibility, solidarity and dignity were presented as being necessary in the age of the genetic revolution in order for humanity to realise its dream of scientific progress and prosperous, peaceful living among different nation-states.[33] The President of the ICJ, Justice Bedjaoui, spoke at length about the challenges facing humanity in the wake of the newfound 'power of man to re-model himself and to recompose – for better or for worse – the family, the marital relationship, the parental or filial relationship, the human group, society, the State, democracy – in short, the social, political and legal organisation of international relationships.'[34]

Recalling the debate on the technological gap between the South and the North in the 1960s, which he said had grown into a chasm, he cautioned against allowing this chasm to grow into an abyss, 'to the extent that one day one will no longer be able to speak of humanity and humanity will cease to be a subject in which every individual can feel involved.'[35] He called for 'resolute solidarity and vigilance', and stressed the role of law, both national and international, in order that 'the global genetic revolution can have some beneficial effects on the underdeveloped world'. Numerous references were made to the concept of the CHM, which was one of the important principles in the Universal Declaration, and the draft instrument was held to be a remarkable contribution to the international governance of the human genome.

The Fourth Session of the IBC was held in Paris on 3–4 October 1996. The three discrete topics discussed at the Fourth Session were food and plant biotechnology and ethics, bioethics and women, and protection of human rights in experimental treatment. Dr Lenoir drew attention to the 'the trail being blazed by international law, which is no longer confined to inter-State relations' and stated that 'international law enables the individual to do more and more to defend his or her fundamental freedoms.'[36] Echoing her ideas, the President of France, in his address at the opening session of the IBC, stated that 'in proposing to raise the human genome to the rank of a common heritage of humanity, you enrich the fundamental rights of individuals and help to ensure respect for the dignity of each human being.'[37] Referring to the 'revolutionary' concept of the CHM in the draft instrument, despite the insertion of the term 'symbolic' to qualify it, the President of the ICJ argued that the concept played a central role in the Universal Declaration. He suggested that it should be given greater value 'by being 'entwined' into the wording of Article 1.[38]

The President of the Legal Commission of the IBC, Hector Espeill, supported the view held by the ICJ President and recalled an earlier General Conference resolution which referred to our genetic heritage as part of the CHM.[39] A biological notion of the CHM was put forward by some members. A biological notion of common heritage is the understanding that the human genome is the CHM based on the fact that all human beings share the genome. From another angle, however, we in fact share our genes with the fruit fly and the banana! We also share 98 per cent of our genes with the chimpanzee. While some acknowledged that the human genome was indeed the CHM, in the biological sense, others disagreed about attaching such a tag to the human genome, since 100 per cent of the genome was not common to all the individuals in the human species.[40] The IBC also discussed mechanisms for implementation of the principles encapsulated in the Universal Declaration.[41]

Following these discussions on the draft instrument to protect the human genome, in early 1997 the IBC recommended the adoption of the Universal Declaration on the Human Genome and Human Rights by the General Conference.

2.2.4 The Universal Declaration on the Human Genome and Human Rights (1997)

The Universal Declaration on the Human Genome and Human Rights forms the basis of 'soft law' in the area of human genome governance. The potential for discrimination arising from the discovery of genetic traits of individuals and ethnic groups was identified as an imminent threat by the IBC and the statements in the Universal Declaration reflect this. Statements against any form of discrimination occupy a central place in the Universal Declaration. Human dignity is reaffirmed under all circumstances and forms the basis of the categorical assertion prohibiting all forms of discrimination based on genetic characteristics. The Universal Declaration rejects a 'doctrine of the inequality of men and races'.[42] The human genome is acknowledged as the heritage of humanity and the foundation both of the diversity and dignity of human beings.[43] And on this understanding there should also be no discrimination caused by unequal benefits arising from the applications of the results of human genomic research, and the benefits should go to 'humankind as a whole'.[44]

The Universal Declaration seeks to attach itself to a framework of human rights[45] based on existing human rights instruments and the enumeration of a number of other personal rights. The core international human rights treaties find mention in the Preamble alongside a number of other treaties and instruments on security and trade. The requirement of a prior, free and informed consent[46] is one of the first of personal rights listed. The principle of non-maleficence – one of the Nuremberg principles, and also a Beauchamp-Childress principle is the source of provisions requiring an assessment of risks and benefits before any procedure is carried out.[47] The right to confidentiality of one's genetic data[48], along with the right to not know the results of a genetic test[49] can only be overridden 'for compelling reasons within the bounds of public international law and the international law of human rights'.[50] Similar constraints are laid on the consent principle too. 'Human rights, fundamental freedoms and human dignity of individuals or, where applicable, of groups of people' are to be respected when research is carried out on the human genome.[51]

Freedom of research, 'a part of freedom of thought'[52], is considered fundamental to the progress of science and technology and the Universal Declaration seeks to ensure that such a freedom is available. States have the twin responsibilities to foster favourable 'intellectual and material conditions' for the exercise of freedom of research and to ensure that ethical, social and legal implications of research are heeded in the process.[53] States are recommended to establish 'independent, multidisciplinary and pluralist ethics committees' in this regard.[54] The Universal Declaration, in Article 12(b), seems to qualify the exercise of the freedom of thought and research with the clause 'The applications of research … shall seek to offer relief from suffering and improve the health of individuals and humankind as a whole.' It is open to debate as to what extent the freedom of research was to be used to respond to social and egalitarian concerns about 'the health of humankind as a whole.' The liberal use of references to 'humankind as a whole' and solidarity would suggest that such freedom was indeed to be used subject to restrictions as stated. Without a direct mention of the right of the health, the Universal Declaration alludes to the benefits of genomic research to the health of individuals and groups.

States are conferred the primary responsibility to protect human rights and freedoms that might be affected by the research on the human genome.[55] Research, training, open debate and vigilance are among the activities States are encouraged to undertake. Only peaceful uses of research results are allowed. A large part of the Universal Declaration focuses on building solidarity for the promotion of the benefits of human genomic research.[56] Beginning with the promotion of solidarity between individuals affected by a genetic condition, the Universal Declaration widens out to solidarity between families and populations,[57] between scientific groups[58] and between ethics committees.[59] Recognising that developing countries would be particularly disadvantaged in ensuring that any benefits of human genomic research would reach their populations, Article 19 makes a series of recommendations for developing a framework of international cooperation which enables risk assessment, capacity building and exchange of scientific knowledge.

IP is the proverbial elephant in the room in the Universal Declaration. It receives a couple of mentions in the Preamble and in Article 4, which states 'The human genome in its natural state shall not give rise to financial gains.' No further elaboration exists on this aspect of research on the human genome.

The principles contained in the Universal Declaration are an attempt by the international community to develop bioethical principles within a framework of human rights and situated within international law. Far from leaving the governance of human genomic research to decentralised or self-regulatory control by practitioner bodies and solely by national legislation, this Universal Declaration emphasises the role that international instruments can play in regulating new technology.[60]

The value of human genetics lies primarily in its capacity to address human disease and help achieve a healthier world by enabling physicians to address disease with an understanding of the genetic factors that cause diseases. To address health via a human rights framework is perhaps to place too much optimism on international human rights law. There are two main issues involved – the recognition of economic social and cultural rights as human rights that need to be enforced and a clearer understanding of what the right to health means in this context. In order to provide a robust human rights framework to regulate research on the human genome it is essential to address and make progress on both the above problems if we are not to be left with a mere collection of personal rights.

Because science had become more expensive and increasingly unaffordable at a private level, State resources were sought for funding science and technology. Therefore the involvement of States in science and technology increased to a large extent and can be seen in the HGP where the State played the leading role in funding and experimentation.[61]

When the UNESCO Director-General convened an expert committee on bioethics in 1993 for the discussion of bioethical issues and the committee began its deliberations, the field was widened to include issues of distributive justice and the deteriorating state of health resulting from epidemics of common diseases. At the global level, the 10/90 gap had been formulated and occupied centre stage in debates on resource allocation and health research.[62] The general feeling was that there were huge inequalities in health between populations in the developing world and the developed world and between rich and poor populations within the developing and developed worlds.

2.2.5 The human genome – subject or object of international law?

One of the themes to emerge from the discussions at the IBC was that of the status of the human genome. The human embryo has been the subject of much debate at national parliaments and among the public. It is most contentious when the debate focuses on human embryo research and the point at which such experimentation should stop and respect for life be ensured.

The main headings extracted from the Universal Declaration on the Human Genome and Human Rights point to three core areas – the status of the human genome, the rights of individuals including researchers and subjects, and the need for solidarity and international cooperation. Human dignity features as the running theme that is used as a justification and rationale for principles put in place to protect human rights.

Section 1 of the Universal Declaration builds an understanding of the term 'human genome'. The likeness and unlikeness among members of the human species are both attributed to the human genome and the information it holds. While our bodily functions are performed because of the information carried within our genomes, the reason why each of us enjoys different levels of health and suffer from different types of diseases is also because there are slight variations of this information that we carry in our genomes. The unity and diversity among the human species is explained by the genome. The genome is influenced by external factors and is subject to change (mutations), making it an entity that is constantly evolving. The significance of the human genome is its function as the repository of codes which are then decoded to enable a person to function. However, the significance of the human genome cannot be properly understood without understanding that it does not work as a standalone factor in human growth and behaviour. The influence that non-genomic factors – the 'natural and social environment'[63] including 'living conditions, nutrition and education'[64] – have is crucial to the way the information in the genome is expressed. Therefore human beings cannot be considered to be the sum total of their genetic characteristics.[65]

The third section of the Universal Declaration focuses on research that is permissible using the human genome. Any research that is contrary to human dignity, human rights and fundamental freedoms shall not be practised and particular emphasis is placed on inclusiveness by considering 'humankind as a whole'[66] with, as 12(a) states, 'due regard for the dignity and human rights of each individual'. Notably there is no restriction on procedures that can be carried out on the human genome so long as they are not contrary to the above stated principles. The International Declaration on Human Genetic Data adopted by UNESCO in 2003 lists various procedures by which the genome can 'be collected, processed, used and stored'. The International Declaration treats biological samples and genetic data distinctly. Biological samples are defined as 'Any sample of biological material (for example blood, skin and bone cells or blood plasma) in which nucleic acids are present and which contains the characteristic genetic make-up of an individual'.[67] Human genetic data refers to 'Information about heritable characteristics of individuals obtained by analysis of nucleic acids or by other scientific analyses.[68] Human genetic data (and also human proteomic data[69]) can be used for the purposes of diagnosis and health care; for medical and other scientific research, including epidemiological and especially population-based genetic studies; for anthropological or archaeological studies, forensic medicine, civil, criminal and other legal proceedings; and for any other purpose consistent with the Universal Declaration and the international law of human rights.[70] The human genome is a resource; it is the raw material from which much information can be extracted to aid myriad types of research ranging from archaeological studies that trace human evolution to the study of diseases to the providing of evidence in civil and criminal proceedings in courts. The genome contains valuable information that is as yet unexplored. In the information age, the human genome is the new natural resource that is the source of knowledge and in turn of power to those that can exploit it.

In Article 1 of the 1997 Universal Declaration, the human genome is given a special status as the 'heritage of humanity'. This is further elaborated in the 2003 International Declaration where Article 4 confers special status on human genetic data because 'they can be predictive of genetic predispositions concerning individuals[71] and they may have a significant impact on the family, including offspring, extending over generations, and in some instances on the whole group to which the person concerned belongs'.[72] The International Declaration also recognises that such data might have special cultural significance for some groups.[73]

Although this special status is given to human genetic data in Article 4, there is also mention of biological samples within Article 4. Throughout the International Declaration human genetic data, human proteomic data and biological samples are referred to as related concepts. This is justified by the fact that biological samples are the carriers of the genetic data and any change or mutation that occurs to the genome is via the physical biological sample.

The purposes for which the genome is being put to use/and could be put to use suggest that it is a natural resource and can be conferred the status of an object in international law. Both the 1997 Universal Declaration and the 2003 International Declaration treat the human genome as a resource that is valuable to scientists. But at the same time, the concept of human dignity is linked to the human genome which 'underlies … the recognition of their (members of the human family) inherent dignity'.[74] Because the genome is the raw material, which under the influence of non-genomic factors works to shape a human person, the human genome is a major part of who we are and who we are constantly becoming. So the natural resource is both animate as well as inanimate, making it a candidate for being counted both as a subject as well as an object in international law.

Under national legislation human embryos are protected to different stages of growth. The human genome and the human embryo are very closely linked. The embryo contains the genetic material (human genome) that grows and develops into a baby. While the embryo is a collection of varying number of cells at different stages of its development, it embarks on its journey as just a single cell whose essence is the human genome.

The status of the human genome is tangentially a human rights issue. States raised the issue of definition at the 1997 General Conference[75] and many commentaries have been written on the moral status of the human embryo[76] but the human embryo is not an entity recognised by international human rights law or within wider international agreements and other instruments.

The international trade rules pertaining to the export and import of human tissue, including the human embryo, treat the embryo as an object. The status of the embryo occupies a central place in bioethical discourse. The rights of a person can sometimes be overridden by the welfare of the embryo, by virtue of its status under a specific legal system.

The status of the embryo is different in different countries.[77] The UK adopts the gradualist approach[78] whereby the embryo gradually acquires a status and until day 14 of its development research which will harm the embryo can be carried out. At the other end of the spectrum are countries which do not allow any research which destroys the embryo. Advances in genomics and genetics has led to the increased use of embryos in research and have hence led to bioethical discussions which have now spilled over into the international arena. The General Assembly, in 2001, discussed an agenda item placed before it by the governments of France and Germany, relating to a ban on reproductive human cloning. The status of the embryo was the decisive point in the debate on the agenda item.

Human cloning is of two types – human reproductive cloning and human therapeutic cloning (also known as research cloning). Cloning is a scientific technique which enables copying of cells and a clone simply means a copy. Skin cells extracted from an individual can be used for cloning research. An egg (ovum) is taken and the nucleus is replaced by the nucleus found within the skin cells, thus resulting in a cell identical to the skin cells – a clone. In this case it produces a cell similar to that of an embryo. With such a cloned cell two possibilities arise: in human reproductive cloning, this cell can be placed within the womb and be carried to term to produce a baby genetically identical to the donor of the skin cells; in therapeutic cloning, the cloned cell is used for research by harvesting stem cells which are a unique kind of cell capable of growing into different types of tissue cells in the body. The status of the embryo is the most contentious point in the debate on human cloning because cloning leads to the destruction or manipulation of the embryo.

Countries differed widely on the issue and the outcome was a divisive vote on the resolution on human cloning adopted by the General Assembly in 2005. The differences arose because of varied positions taken by States on whether the embryo was 'life' or not. While Italy, Spain, the US and other States granted full personhood to the embryo because of their position that an embryo represented life in the same way as a human being represented life, countries such as Britain were reluctant to go along with that position because they were of the opinion that the status of an embryo was lower than that of a human being.

The Convention on the Rights of the Child (CRC),[79] which came into force in September 1990, states 'the child, by reason of his physical and mental immaturity' needs 'special safeguards and care, including appropriate legal protection, before as well as after birth'. But the reference here is not directly to an embryo or even to any particular stage in foetal development. In juxtaposition to the rights of the child, many countries allow for the destruction of the foetus before birth, setting various limits on the time period before which such destruction should take place. Abortion laws in countries vary but it is notable that a foetus can be aborted up to 168–196 days after conception.[80] In recent years premature babies born much before this period have survived,[81] thus initiating calls for a review of abortion time limits. Although the CRC drafters although found it necessary to include the period of the birth in their consideration of the rights of the child, they intentionally avoided setting limits on the length of time before birth that such rights extend to. It cannot be presumed that such an omission means the whole length of time before birth from the date of conception because the CRC supports abortion when it is legal under any particular jurisdiction. The monitoring committee also urged States to provide for safe abortion services for adolescent girls.[82] It was argued that the Universal Declaration of Human Rights and the *International Covenant on Civil and Political Rights* explicitly recognise rights only for persons who are born and attempts to specify 'the moment of conception' were rejected.[83]

Michel Meslin concludes that the concept of person is neither a simple fact nor evident throughout history.[84] The concept is difficult to define with certainty based on the already established rules and practice in international law as to whether the human genome should be construed as a subject or object in international law. Despite the unique nature of the genome and the special status that genetic data is accorded, the closer the entity/foetus/embryo gets to its formation the less special that status seems to be. There is a corresponding sliding scale of protection according to how developed the entity/foetus/embryo is. While the foetus is protected by certain rights in most countries, it seems that even countries that do now allow abortion still allow some amount of research to take place on the embryos at the very early stages. And in assisted reproduction procedures 'surplus' or 'waste' embryos are regularly discarded as is the wastage that occurs when preservation methods fail.

The above discussion has shown that the human genome is a special natural resource[85] that shapes and affects humanity in a fundamental way, and that it should thus be subject to international governance in a fair and equitable manner. Natural resources have been exploited to the detriment of various groups and populations throughout history and the danger that history would repeat itself is very real in the context of the human genome.[86] The 1997 Universal Declaration, in Article 12(a), suggests that 'Benefits from advances in biology, genetics and medicine, concerning the human genome, shall be made available to all'. In elaborating this provision further in Article 12(b) it is stated that 'applications in biology, genetics and medicine, concerning the human genome, shall seek to offer relief from suffering and improve the health of individuals and humankind as a whole'. Article 12 refers to the health of individuals and not explicitly to the right the health. This is more clearly stated in the Universal Declaration on Bioethics and Human Rights adopted by UNESCO in 2005, Article 14(2) of which states that the 'enjoyment of the highest attainable standard of health' is accepted as 'one of the fundamental rights of every human being without distinction of race, religion, political belief, economic or social condition'. The recognition that the human rights framework should play an active role in achieving international bioethical goals should be understood as part and parcel of a larger framework of securing equitable benefit for 'humankind as a whole'. Article 19 of the 2005 Universal Declaration sketches out a set of guidelines with the principle of solidarity as the basis of cooperation between individuals, families, population groups and among developing and developed countries with the objective of 'economic and social progress for all'.[87] The concept of the CHM has had a role to play in the IBC discussions and can be considered as the framework concept for the international governance of the human genome. This will be explored in the later chapters of this book.

The 1997 Universal Declaration read along with the 2005 Universal Declaration deepens the understanding of economic social and cultural rights. The 1997 Universal Declaration sketched a framework for human genome governance which was further strengthened by the 2005 Universal Declaration which elaborates on provisions relating to solidarity and health.

2.3 The UN and the human genome: the UN Commission on Human Rights (UNCHR)

The UNCHR[88] added to its agenda the issue of human rights and bioethics in 1993. In its resolution dated 10 March 1993 the UNCHR recognised the need to preserve the dignity and integrity of the human being in the wake of scientific progress and referred to the two international covenants on human rights as setting the standards for protection.[89] The UNCHR has now been replaced by the Human Rights Council (HRC) as part of the UN reform process.[90] The HRC reports directly to the UN General Assembly unlike the UNCHR which was one of the subsidiary bodies of the Economic and Social Council of the UN (ECOSOC). The UNCHR was aided in its work by the Sub-Commission on the Promotion and Protection of Human Rights[91] which acted as a 'think tank' for the UNCHR. It played an advisory role and assisted in bringing numerous issues to the UNCHR's attention.

The UNCHR did not put much effort into examining the impact on human rights caused by research into the human genome or the subsequent applications of such research. Generally however, the resolutions on science and technology can be taken to cover the human genome. Specific references begin to appear from 1997 onwards following the initiative of UNESCO representatives present at the UNCHR's meetings. In its endeavour to ensure that scientific knowledge respected the general principles enshrined in the Universal Declaration of Human Rights, the representative of UNESCO presented the 1997 Universal Declaration as an important 'move' on the part of her organisation that was 'designed to protect the dignity and indivisibility of the human person and the human family by seeking to ensure that the human genome was not exploited for commercial purposes'.[92] Following this, the UNCHR seemed to defer to the role played by UNESCO in regard to the issue of bioethics and almost ignored human rights aspects when it alluded to 'the ethical, social and human questions raised by the biomedical research'[93] and not to human rights per se in its discussions. It further invited governments 'to consider establishing independent, multidisciplinary and pluralist committees of ethics to assess, notably in conjunction with the International Bioethics Committee of the United Nations Educational, Scientific and Cultural Organization, the ethical, social and human questions raised by the biomedical research undergone by human beings'.[94] It however tasked the Sub-Commission (which was known as the Sub-Commission on Prevention of Discrimination and Protection of Minorities until 1999) to consider the nature of the contribution that could be made to 'the reflections of the International Bioethics Committee'.[95] This continued for a few years[96] until the Sub-Commission became actively involved in putting a proposal forward.[97] The Sub-Commission expert Professor Motoc (Romania) addressed the importance of understanding the complexity of how to deal with such research. She pointed to the balance that needed to be achieved between the freedom to research and the need to ensure that scientific research protected the interests of vulnerable subjects.[98] Her recommendation to the Sub-Commission was appropriately broad, in that a follow-up of the adoption of the 1997 Universal Declaration be pursued by the Sub-Commission.

Professor Motoc was charged with leading the follow-up on the 1997 Universal Declaration and produced a series of reports between 2002 and 2006, each building on her previous reports. The themes covered were very similar to those covered by UNESCO. There was strong concern, particularly among some members from developing countries, that the knowledge and application of human genome research would stay within certain pockets in affluent countries. There was also widespread apprehension that knowledge acquired via the study of the human genome would affect disability rights and bring in discriminatory practices.[99] The notion that the human genome was the heritage of all humankind figured frequently in the debate, both as rhetoric and in juxtaposition with intellectual property rights (IPR).[100] The involvement of private commercial interests in genome research was 'a great concern',[101] seen as having a great impact on access to the benefits of advances in science and technology related to the human genome.[102] The rights of indigenous people to control the use and commercialisation of their own genes and the plant biodiversity that they help foster would be in danger.[103] Cloning and its human rights implications formed part of many statements made by the experts, although without much elaboration on whether therapeutic cloning or reproductive cloning was being referred to and what the different ethical positions were with respect to this procedure, which is quite different to the human genome. The lead for this was perhaps taken from Article 11 of the 1997 Universal Declaration which refers to reproductive cloning as an example of a practice that should prohibited on the basis that it violated human dignity.

2.3.1 The expanded working paper

In 2003 an expanded working paper was submitted to the Sub-Commission[104] by Professor Motoc who was appointed the Special Rapporteur on the subject of protection and fulfilment of human rights in the course of research on the human genome. The paper lays out an agenda for the Sub-Commission to consider. Discrimination, the ethics of genetic manipulation, the CHM concept, and IPR in the human genome figure in the expanded paper. Clear objectives emerge from the section on discrimination concerning the potential for discriminatory treatment of individuals by their employers or insurers based on results obtained by genetic tests which identify susceptibility to genetic disorders. Responding to the potential of such discrimination, the paper recommends, and in the process reiterates, that 'Discrimination against a person on the ground of his/her genetic heritage should be expressly prohibited. Tests and therapies would not be used in insurance or employment contexts. Rules should also be established to govern the disclosure of genetic information to third parties.'[105] The discussion on other issues makes comprehensive reading on the complexities and challenges faced by human rights regimes in order to ensure the protection of dignity and rights of individuals, and the author reckons that human genome research has necessitated the law to move into uncharted territory.[106] In the sections elaborating the three other main themes the author has chosen to present, namely (1) The Human Genome: Common Heritage of Humanity; (2) Intellectual Property Rights and the Human Genome;

and (3) Human Genetic Manipulation and Human Rights, general pointers on the human rights issues that are to be taken up by the Sub-Commission are not very apparent.

The tension between developed and developing countries on the CHM regime and its applications in various international treaties attracts this statement from the author with respect to how the debate on the CHM concept should influence an international framework on the human genome: 'An international regime concerning the human genome must be interpreted so as to assure equilibrium between the needs of both developed and developing countries'. Clearly there have been efforts to draw out human rights statements on development that reflect on the need to secure to the individual their corresponding 'development right'. A working paper on the right to development was being prepared almost in parallel by another Sub-Commission expert.[107]

Genetic manipulation and germ-line therapy would have been the expected subjects of analysis under the section entitled 'Human Genetic Manipulation and Human Rights' but the focus is instead on cloning. Genetic manipulation combined with human reproductive cloning would be a potent combination which would be ethically and legally questionable; however mere cloning does not amount to the genetic manipulation of the individual, although the argument can be advanced that if done on a large scale it could affect the diversity of the gene pool. In the words of James Watson, Nobel Laureate and former director of the Cold Spring Harbour laboratory, where the plans were drawn up for the HGP:

> our research on DNA has had no impact on cloning. Cloning could have been done without knowing about the structure of DNA. It's not a moral dilemma created by DNA research, but it's created by understanding the biology of human reproduction better. Human cloning is something that really became possible as a result of techniques developed for in vitro fertilization.

Such confusion is found even in national legislation and points to the fact that the science behind the new developments blurs boundaries in many different ways and that definitions of genetic manipulation and the genome have not yet been fixed.[108]

The section of the working paper on 'Intellectual property rights and the human genome' identifies many issues and mentions the European Patent Convention (1973),[109]the Biotechnology Directive[110] and the Agreement on Trade-Related Aspects of Intellectual Property Rights (TRIPS).[111] The patentability of human gene sequences as a human rights issue is just beneath the surface of the discussion, except when the author brings to the notice of the members Sub-Commission Resolution 2000/7 where the conflict between IPR and human rights is highlighted.[112] Another issue mentioned very briefly calls into question the application of IPR to genetics and the clash between such rights and the concept of traditional knowledge, where such knowledge is considered as not being susceptible to IP regimes.[113]

Overall the expanded working paper informs the Sub-Commission of the terms of debate on the governance of the human genome and is intended to cover the whole range of issues on the horizon, which it does well. However, pressing issues such as the nature of the rights investigated in the context of the research on the human genome and its applications have not been considered. The reports submitted by Professor Motoc bring no unique human rights perspective to the discussion. Philip Alston, the chairperson of the Committee on Economic, Social and Cultural Rights criticised UNESCO for adopting the '*so-called* "Universal" Declaration on the Human Genome',[114] and said that the present opportunity created by the Sub-Committee could have been utilised for addressing issues concerning the nature of rights in the document adopted in 1997 by the General Conference of UNESCO (emphasis added). Whether human rights should play a role in global governance of the human genome is an important question that States face.[115] On the one hand there are doctrinal issues about whether the rights listed in the 1997 Universal Declaration are indeed human rights and, if they are, what type they might be.[116] And on the other hand, there are many culturally sensitive aspects of human rights which could be accentuated in the context of using a human rights approach to regulation of human genomic research.

In the next two reports that followed, in 2004[117] and 2005,[118] Professor Motoc dealt with the subjects of genetic information, the rights surrounding it – including the right to privacy and the right to not be discriminated against on account of it –, and the principles of IPR surrounding genetic information.

2.3.2 The preliminary and interim reports

The preliminary report of 2004 adopts the understanding of genetic discrimination to be 'the differential treatment of individuals or their families based on their actual or presumed genetic differences (social discrimination) as distinguished from discrimination based on having symptoms of a genetic-based disease (medical discrimination).' The dilemma remains as to whether new laws on genetic discrimination would be counterproductive in that they would reinforce the idea of genetic determinism, that the sum total of one's genes determines one's individuality.[119] When a State protects its people against all forms of discrimination via constitutional arrangements, criminal laws, data protection, human rights and other legislation, this would seem sufficient to cover against any breaches. The inherent difficulty in ensuring privacy of genomic information when such information is shared with family members is discussed as a challenge of individual and group consent in the next section of the report. The potential misuse of genetic information to curtail benefits and to practise discrimination by employers and insurance companies is presented as a continuing challenge to human rights. The interesting section on indigenous groups and how the rights to privacy and concepts of identity are challenged by such groups brings out the complexity of international governance in the area of human genomic research. A fixed understanding of privacy and identity could potentially compromise the right to self-determination and the cultural independence of indigenous peoples.[120] Benefit sharing by providing access to medical care, capacity building for research purposes, and other measures are supported in the report.

Unlike UNESCO's IBC, Professor Motoc does not hesitate to address the issue of IPR in her interim report published in 2005,[121] perhaps because of the different approaches adopted – 'the ethics approach taken at UNESCO and the legal one assigned to the Office of the High Commissioner'.[122] The interim report draws attention to Article 27(2) of TRIPS 'provides for the possible exclusion from patentability of certain innovations in order to protect human, animal or plant life or health other than micro-organisms, and essentially biological processes for the production of plants and animals other than non-biological and microbiological processes.'[123] By also citing Article 31 of TRIPS which allows for compulsory licensing of inventions, Professor Motoc was seeking to highlight some of the flexibilities available within the system.

One possible approach to IP issues is to probe into the nature of IPR and the rationale behind the granting of such rights, and the nature of property rights as human rights. The right to property characterised as a bundle of rights with various actors holding varying degrees of control over a thing can also be applied to IP. An analysis based on the nature of property rights in IP regimes would open new possibilities for the IP governance of human genome research. The interim report refers to the UN Development Programme (UNDP) which promotes the initiative to 'begin dialogues to replace TRIPS … with alternate intellectual property paradigms'[124] because the organisation is uncertain of the relevance of a global IP regime such as TRIPS for developing countries. Adding weight to these doubts are similar arguments about whether the globalisation of IPR protection will be a socially optimal system, although it might be an IP owner's dream. If IPR is conceived as a form of tax paid by consumers of knowledge to producers of knowledge, the tax should be set at the optimal level, a level high enough to equal 'the social benefits of the knowledge that would not otherwise be produced.'[125] In other words 'just as too low a tax gives us unduly low amounts of innovative incentive, so too does too high a tax represent an unnecessary and socially wasteful transfer from consumer to producers.'[126] A global IP system should rethink its founding basis, otherwise the individual producers and innovators will be benefiting disproportionately compared to the global consumers. While the rights and interests of the producers seem adequately protected in such an arrangement, it is difficult to even identify what the rights and interests are of the benefiting consumers. The WHO has already seen fit to proceed with new approaches to innovation in pharmaceutical research; the focus in this approach is the protection of the interests of consumers to enhance their access to medicines.[127] The ideal goal of an IPR system should indeed be the marriage between the TRIPS market-driven research and development approach and the WHO health needs-driven research and development approach.

Tracing the history of plant genetic resources and their regulation in international law, the report addresses the usefulness of the CHM concept for the debate on human genome governance. It concludes that given the direction taken by the international community by agreeing upon sovereign rights over plant genetic resources, and seeking to protect indigenous rights in forums such as the WTO, a CHM regime for the human genome would only 'prevent scientific research from developing'. Professor Motoc calls for a reconciling of private and public interests.

The CHM concept is central to the following principles when applied to the human genome, according to the expanded working paper submitted in 2003:

i the genome could not be appropriated by any country or private corporation within that country;
ii all States would share responsibility for setting regulations and laws for permissible uses of the genome;
iii all States would share in the benefits derived from HGP, which would mean that all gene sequences would be publicly accessible;
iv the genome would be reserved exclusively for peaceful uses;
v the international community would have shared responsibility for preserving the genome intact for future generations.[128]

The paper makes an assumption that existing models of CHM, which are more suitable for natural resources such as the deep seabed minerals, can be applied without much modification to the human genome, which is a special natural resource. Although the 1997 Universal Declaration suggests a framework similar to that of existing models, there are other human rights provisions which need to be considered in the light of the special status of the human genome and which are also contained in the 1997 Universal Declaration. The structure of such a CHM regime would be different to that of the existing models, while retaining the core principle – which is to benefit the development of each individual, and ultimately to benefit humanity as a whole. The dismissal of the concept of the CHM in the expanded working paper and in the interim report is premature.

Pending the final report, which is to be submitted to the newly formed Advisory Committee, it remains to be seen if the HRC would continue in its engagement with the governance of the human genome.

2.4 The World Health Organization (WHO)

The involvement of WHO in the regulation of public health, health research and health care has been minimal to date. In 2003 the first international agreement orchestrated by WHO, the Framework Convention on Tobacco Control (FCTC)[129] was opened for signature. FCTC is an evidence-based robust regulatory strategy developed to control the demand and supply of tobacco, reaffirming the right of all people to the highest standard of health. It came into force in February 2005 and has 160 ratifications. The FCTC does not overtly adopt a rights-based approach to fulfil its objectives but the individual is its focus and there are references to Article 12 of *International Covenant on Economic, Social and Cultural Rights* (ICESCR), the CRC and the protection of women and vulnerable groups. More recently, after the panic caused by the rapid spread of the SARS virus in 2005, WHO undertook the revision of International Health Regulations (IHR) to strengthen global public health security. The revised regulations came into force in June 2007 and require Member States to take a set of measures to respond to emergency public health threats.[130] Historically the agency has chosen to eschew the legal approach[131] in favour of developing international guidelines of practice for specific health issues. But its involvement in health promotion has been extensive across different parts of the world and the World Health Assembly (WHA) has adopted numerous resolutions over the decades including the Health for All agenda, which continues to play an important role in governance debates on health and research. In addition to this, WHO's policy and practice in developing health strategies for low- and middle-income countries and in addressing issues relating to health equity and access can form the basis for shaping the human right to health and setting its parameters.

WHO's Human Genetics Programme (HGN) helps developing countries with the realisation of their goals in the prevention and management of genetic disorders. The WHO acknowledges the 'unparalleled opportunity to advance the health of all'[132] provided by the rapid growth of human genome research and its acceptance into medical practice, but has reported that States have to take measures to ensure that such benefits are equitably distributed. Its ELSI programme aims to promote equitable access to health benefits such as genetic tests, to focus on genomic health research prioritisation in order to benefit developing countries, and to build capacity in human genomic research. A report published in 2002 encouraged 'Member States to develop clear and concise ethical frameworks to guide the conduct of genomics research and its medical application in conjunction with the social, economic and religious context unique to their countries.' Global collaboration to identify universal codes of practice was however favoured in order to provide the basis for such national ethical frameworks.

WHO's contribution to the governance of genomic research has been through its efforts to link trade and health issues. The FCTC is an example of the breadth and scope of cross-cutting measures that need to be addressed when dealing with health. The WHA passed a resolution on IPR, innovation and public health in 2003 highlighting concerns about the outcomes of the current patent protection systems with regards to public health. IPR are in themselves rights that offer protection to products of the intellect, but by providing monopoly over such products for certain lengths of time, this system of granting property rights plays a significant role in industry and trade, especially in the global knowledge-based economic system. The 2003 Resolution identified the need to channel innovation in the direction of promoting global health. Under the current systems of incentivising innovators, the resolution identified that insufficient research and development in 'so-called "neglected diseases" and "poverty-related diseases"'[133] was causing inequities in health research. The statistics relied on point to neglect of research into diseases that predominantly affect tropical countries (only 13 out 1,400 drugs developed over a period of 24 years were for tropical diseases). It urges Member States to take measures to protect public health by availing themselves of the flexibilities provided in the TRIPS Agreement. By reiterating that 'Member States should urge the pharmaceutical industry to reinvigorate its efforts to develop innovations that add real therapeutic advantage in treating the world's major killer diseases, especially in developing countries', the WHA expressed reservations about the current system's suitability in responding to health innovations.

The provisions on patent in TRIPS are most relevant to health. Under

Article 27(2) State parties are allowed to exclude the granting of patent protection to inventions that are deemed necessary to protect public health, and the same applies to diagnostic, therapeutic and surgical methods for the treatment of humans (Article 27(3)(a)). The resolution supported the use of compulsory licensing mechanisms to relax patent protection where it was necessary to do so. It also recognised that compulsory licensing mechanisms would not still achieve health goals where States had limited or no manufacturing capabilities, and urged Member States of the WTO to address this issue.[134] Cooperation between Member States and between developing and developed countries to exchange and transfer technology and research findings is recommended as a means to address public health problems.[135]

Property rights in the human genome have been problematic.[136] Along with the set of issues already discussed in relation to property rights, the resolution of the WHO and the report on which it draws bring out further inadequacies in the IPR system and the mechanisms that have evolved around it. The earlier discussion revealed that although an individual cannot benefit from their own tissue samples, property rights can be obtained by others who invent a product from it. From the perspective of the inventor this is a favourable situation and one that helps satisfy the objective of such a system which is to promote innovation. On the other hand, the above discussions have identified some further issues with respect to the granting of IPR and the disincentives it provides for innovation in low-return health R&D. This raises serious issues about whether IP system, and particularly the system of patents, can fulfil the 'other' purpose of its existence, which is to bring benefits to society. Health goals are among the important benefits that society can gain from at the current time as identified by the UN General Assembly's Millennium Declaration.[137] Following the report of an Intergovernmental Working Group on Public Health, Innovation and Intellectual Property, Member States adopted a global strategy in May 2008 to address innovation in public health. The strategy is a clear indicator of the gap left in innovation because of the current dominant approaches to innovation. Instead of focusing on the extent and nature of the protection granted to innovators, underpinned by the philosophy that the stronger the protection the more beneficial it is to society, the WHO global strategy seeks to identify the benefits that society requires in order to achieve its goals and to stimulate invention in that direction.

Notes

1 See R. J. D. Macer, 'Revised Outline of the (UNESCO) Declaration on the Protection of the Human Genome', *Eubios Journal of Asian and International Bioethics* 5, 1995, 97–9.

2 See Article 1, Universal Declaration on the Human Genome and Human Rights, UN Doc. A/RES/53/152, 1999.

3 The Workshop on International Co-operation for the Human Genome Project, 24–26 October 1988. See S. Grisolia, Mapping the Human Genome, *The Hastings Center Report* 19 (4), 1989, 18–19.

4 Council for International Organizations of Medical Sciences (CIOMS), reprinted

in A.R. Jonsen, *et al.* (eds), *Source Book in Bioethics: A Documentary History*, Georgetown: Georgetown University Press, 1998, p. 322.

5 Article VII of the Declaration of Inuyama states 'the needs of developing countries should receive special attention to ensure that they obtain their due share of the benefits that ensue from the human genome project. In particular, methods and techniques of testing and therapy that are affordable and easily accessible to the populations of such countries should be developed and disseminated whenever possible.' The WHO gave its assessment of the technology when its Advisory Committee on Health Research published the report *Genomics and World Health* in 2002 that evaluates the impact and potential of genomics on world health. Its conclusions express strong reservations about the benefits that genomic technology and medicine could bring to health care among the populations in its 191 Member States.

6 See 'Ethics', Paris: UNESCO, available online at http://www.unesco.org/ethics (accessed 21 May 2006).

7 See 'International Bioethics Committee', Paris: UNESCO, available online at http://portal.unesco.org/shs (accessed 21 May 2006).

8 Ibid.

9 Ibid.

10 UNESCO Doc. C/RES/32/22, 2003.

11 Adopted by acclamation on 19 October 2005 by the 33rd Session of the General Conference of UNESCO.

12 See COMEST, Paris: UNESCO, available online at http://portal.unesco.org/shs (accessed 21 May 2006).

13 Correspondence with Caroline Munier, UNESCO, Paris, 2006.

14 See 'Avicenna Prize for Ethics in Science', Paris: UNESCO, available online at http://portal.unesco.org/shs (accessed 21 May 2006).

15 See 'International Bioethics Committee', Paris: UNESCO, available online at http://portal.unesco.org/shs (accessed 21 May 2006).

16 P. Sands and P. Klein, *Bowett's Law of International Institutions*, 5th edn, London: Sweet and Maxwell, 2001, p. 286.

17 World Medical Association, Declaration on the Human Genome Project. 1992. Adopted by the 44th World Medical Assembly, Marbella, Spain, September 1992, available online at http://www1.umn.edu/humanrts/instree/humangenome.html (accessed 3 July 2006).

18 *Proceedings of the First Session of the IBC*, Paris: UNESCO, 1993, available online at http://portal.unesco.org/shs (accessed 15 May 2006).

19 See UNESCO Doc. C/RES/28/2.2, 1995, UNESCO Doc. C/RES/22/13.1, 1983, UNESCO Doc. C/RES/23/13.1, 1985, UNESCO Doc. C/RES/24/13.1, 1987, UNESCO Doc. C/RES/25/5.2 and 7.3, 1989, UNESCO Doc. C/RES/27/5.15, 1991, UNESCO Doc. C/RES/28/0.12, 2.1 and 2.2, 1993.

20 *Proceedings of the First Session of the IBC*, Paris: UNESCO, 1993, available online at http://portal.unesco.org/shs (accessed 15 May 2006), p. 59.

21 Ibid., p. 57.

22 Ibid.

23 Ibid.

24 Ibid., p. 61.

25 *Proceedings of the Second Session of the IBC*, Paris: UNESCO,1994, available online at http://portal.unesco.org/shs> (accessed 15 May 2006).

26 Ibid., p. 67.

27 Ibid., p. 69.

28 Ibid., p. 106.

29 *Proceedings of the Third Session of the IBC*, Paris: UNESCO, 1995, available online at http://portal.unesco.org/shs (accessed 15 May 2006), p. 82.

30 Ibid., pp. 97–8.

31 Ibid. See also a paper by the Chairman of the Legal Commission, 'Project of an International Instrument for the Protection of the Human Genome', Paris: UNESCO, 2003, available online at http://www.unesco.org/ibc/en/genome/juridique/instrument.html (accessed 24 November 2001), p. 10.
32 *Proceedings of the Third Session of the IBC*, Paris: UNESCO, 1995, available online at http://portal.unesco.org/shs (accessed 15 May 2006).
33 Ibid., p. 108.
34 Ibid., p. 139.
35 Ibid., p. 142.
36 *Proceedings of the Fourth Session of the IBC*, Paris: UNESCO, 1996, available online at http://portal.unesco.org/shs (accessed 15 May 2006), p. 106.
37 Ibid., p. 112.
38 Ibid., p. 47.
39 UNESCO Doc. C/RES/28/0.12, 1995, on the medium-term strategy of the organisation, unanimously adopted in 1995.
40 *Proceedings of the Fourth Session of the IBC*, Paris: UNESCO, 1996, available online at http://portal.unesco.org/shs (accessed 15 May 2006), p. 48.
41 Ibid., p. 73.
42 Preamble, 1997 Universal Declaration.
43 Article 1, 1997 Univesal Declaration.
44 Preamble, 1997 Universal Declaration.
45 Article 25 states 'Nothing in this Declaration may be interpreted as implying for any state, group or person any claim to engage in any activity or to perform any act contrary to human rights and fundamental freedoms, including the principles set out in this Declaration.'
46 Article 5(b), 1997 Universal Declaration.
47 Article 5(a)(e)(d), 1997 Universal Declaration.
48 Article 7, 1997 Universal Declaration.
49 Article 5(c), 1997 Universal Declaration.
50 Article 9, 1997 Universal Declaration.
51 Article 10, 1997 Universal Declaration.
52 Article 12(b), 1997 Universal Declaration.
53 Article 14, 1997 Universal Declaration.
54 Article 16, 1997 Universal Declaration.
55 Articles 14–23, 1997 Universal Declaration.
56 Section E: Solidarity and International Co-operation, Articles 17–19, 1997 Universal Declaration.
57 Article 17, 1997 Universal Declaration.
58 Article 18, 1997 Universal Declaration.
59 Article 23, 1997 Universal Declaration.
60 S. Harmon, 'The Significance of UNESCO's Universal Declaration on the Human Genome and Human Rights', *SCRIPTed* 20 (2:1), 2005, available online at http://www.law.ed.ac.uk/ahrc/script-ed/vol2-1/harmon.asp (accessed 7 June 2006).
61 It was funded by the National Institutes of Health (NIH) and the Department of Energy in the US, and other countries' genome work was also funded to varying extents from the public purse. See R. Cook-Deegan, *The Gene Wars: Science, Politics and the Human Genome*, New York: W. W. Norton and Company, 1994; R. Cook-Deegan, 'Origins Of The Human Genome Project', *Risk: Health, Safety & Environment*, 1994, 97–118.
62 Commission on Health Research for Development, *Health Research: Essential Link to Equity in Development*, New York: Oxford University Press, 1990; G. Yamey, 'The world's most neglected diseases – Ignored by the pharmaceutical industry and by public-private partnerships', *British Medical Journal* 325, 2002, 176–7.
63 Article 3, 1997 Universal Declaration.

64 Ibid.
65 Also see Article 3 of the International Declaration on Human Genetic Data adopted by UNESCO on 16 October 2003 which describes a person's identity: 'Each individual has a characteristic genetic make-up. Nevertheless, a person's identity should not be reduced to genetic characteristics, since it involves complex educational, environmental and personal factors and emotional, social, spiritual and cultural bonds with others and implies a dimension of freedom.' It brings out the layers of factors involved in shaping a person's identity.
66 Article 12(b), 1997 Universal Declaration.
67 Article 2(i), UNESCO Doc. C/RES/32/22. 2003.
68 Article 2(iv), UNESCO Doc. C/RES/32/22, 2003.
69 Human Proteomic Data refers to information pertaining to an individual's proteins including their expression, modification and interaction (Article 2(ii)). Genes code for proteins which form the building blocks for human tissue, therefore human genetic and proteomic data are linked. By the time the 2003 International Declaration was adopted study of the human proteome (sum total of the proteins) had become more extensive and this was taken into account in the Declaration.
70 Article 5, UNESCO Doc. C/RES/32/22, 2003.
71 Article 4 (a)(i), UNESCO Doc. C/RES/32/22, 2003.
72 Article 4 (a)(ii), UNESCO Doc. C/RES/32/22, 2003.
73 Article 4 (a)(iv), UNESCO Doc. C/RES/32/22, 2003.
74 Article 1, 1997 Universal Declaration.
75 Draft Report of Commission III, UNESCO Doc. C/RES/29/82, 1997, p. 27.
76 W. P. Cheshire, 'Human Embryo Research and the Language of Moral Uncertainty', *The American Journal of Bioethics* 4 (1), 2004, 1–5.
77 See E. P. George, 'The Stem Cell Debate: The Legal, Political and Ethical Issues Surrounding Federal Funding of Scientific Research on Human Embryos', *Albany Law Journal of Science & Technology*, 12 (2), 2002, 747–800; R. M. Isasi, *et al.*, 'Legal and Ethical Approaches to Stem Cell and Cloning Research: A Comparative Analysis of Policies in Latin America, Asia, and Africa', *Journal of Law, Medicine & Ethics* 32, 2004, 626–37; S. Fiandaca, 'In Vitro Fertilization and Embryos: The Need for International Guidelines', *Albany Law Journal of Science & Technology* 8, 1998, 337–404; L. B. Moses, 'Understanding Legal Responses To Technological Change: The Example Of In Vitro Fertilization', *Minnesota Journal of Law, Science & Technology*, 6, 2006, 505–618 at 518; G. Kanellopoulos, 'Embryonic Stem Cell Research: A Comparative Study of the Philosophies of the United States and the United Kingdom', *Journal of International Business and Law* 4, 2005, 170–91; J. D. Ohlin, 'Is The Concept Of The Person Necessary For Human Rights?', *Columbia Law Review* 105, 2005, 209–49; R. J. Araujo, 'The Meaning of Person in the Context of Human Embryonic Cloning – Evolving Challenges for the Rule of Law in the International Order', *University of Saint Thomas Law Journal* 1, 2003, 39–55.
78 *Report of the Committee of Inquiry into Human Fertilisation and Embryology, Cm 9314 (Warnock Report)*, London: HMSO, 1984; M. Warnock (ed.), *A Question of Life: the Warnock report on human fertilisation and embryology*, Oxford: Blackwell Publishers, 1985.
79 Annex, UN Doc. A/RES/44/25, 1989, entered into force 2 September 1990 and reproduced in *International Legal Materials* 28, 1989, 1448.
80 In the UK, termination of pregnancy is allowed for up to 24 weeks (168 days) as per section 37 of the Human Fertilisation and Embryology Act, 1990, which amended the provisions of the Abortion Act, 1967. In Cyprus, abortion is permitted within 28 weeks (196) days. A comprehensive coverage of laws is found in Abortion Policies: A Global Review, *United Nations Department of Economic and Social Affairs. Population Division, United Nations Publications*, 2001; Also see for a comparison between jurisdictions A. Plomer, A foetal right to life? The case of Vo v France, *Human Rights Law*

Review, 5(2), *2005*, 311–38, pp. 334–7; For a more medical discussion of issues concerned see, 'Fetuses and Newborn Infants at the Threshold of Viability A Framework for Practice', *British Association of Perinatal Medicine (BAPM)* Memorandum http://www.bapm.org/documents/publications/threshold.pdf (accessed 27th November 2008)

81 'Baby born after just 22 weeks leaves ward', *Guardian*, February 20, 2007, available online at http://www.guardian.co.uk/world/2007/feb/20/usa (accessed 25 August 2008).

82 Committee on the Rights of the Child, General Comment No. 4: Adolescent health and development in the context of the Convention on the Rights of the Child, in Compilation of General Comments, supra n. 32 at 328, para. 27.

83 C. C. Zampas, J. M. Gher, 'Abortion as a human right: international and regional standards', *Human Rights Law Review* 8 (2), 2008, 249–94 at 263.

84 M. Meslin, 'Religious traditions and the human person', in H. G. Kippenberg, Y. B. Kuiper, A. F. Sanders (eds), *Concepts of Person in Religion and Thought*, Berlin: Mouton de Gruyter, 1990, p. 67.

85 There is no definition available for a natural resource in international law. Some multilateral treaties provide guidance on an understanding of the concept. The Convention for the Protection of the Natural Resources and Environment of the South Pacific Region (Noumea, 24 November 1986) does not define natural resources but includes minerals, wild flora and fauna. According to Article of V the African Convention on the Conservation of Nature and Natural Resources, 'natural resources means renewable resources, tangible and non-tangible including soil, water, flora and fauna and non-renewable resources'.

86 The WHO report, *Genomics and World Health* (2003) warns of unequal benefits deriving from the knowledge acquired from the human genome.

87 Article 19(a)(iii), 1997 Universal Declaration.

88 Established by ECOSOC Resolution 5(I) of 16 February 1946.

89 Commission on Human Rights Doc. RES/67/91, 1993.

90 F. J. Hampson, 'An Overview of the Reform of the UN Human Rights Machinery', *Human Rights Law Review* 7 (1), 2007, 7–27; UN Doc. A/RES/60/251, 2006.

91 Established by ECOSOC Resolution 9(II) of 21 June 1946.

92 ECOSOC Doc. E/C.12/1997/SR.27, 1997.

93 Commission Doc. HR/CN/99/62, 1999.

94 Ibid.

95 Ibid.

96 See Press Releases, Commission on Human Rights, 57th Session, 25 April 2001; Commission on Human Rights, 57th Session, 27 April 2001; and ECOSOC, 24 July 2001.

97 Sub-Commission on the Promotion and Protection of Human Rights, 53rd Session, 15 August 2001.

98 Ibid.

99 Françoise Jane Hampson, Sub-Commission Expert, Sub-Commission on the Promotion and Protection of Human Rights, 54th Session, 15 August 2002.

100 El Hadji Guisse and Antoanella-Iulia Motoc, Sub-Commission Experts, Sub-Commission on the Promotion and Protection of Human Rights, 54th Session, 15 August 2002.

101 David Weissbrodt, Sub-Commission Expert, Sub-Commission on the Promotion and Protection of Human Rights, 54th Session, 15 August 2002.

102 Asbjorn Eide, Sub-Commission Expert, Sub-Commission on the Promotion and Protection of Human Rights, 54th Session, 15 August 2002.

103 Ronald Barnes, of the Indigenous World Association, Sub-Commission on the Promotion and Protection of Human Rights, 54th Session, 15 August 2002.

104 'Human Rights and Bioethics', expanded working paper submitted by Ms. Iulia-Antoanella Motoc in accordance with Sub-Commission decision 2002/114, UN Doc. E/CN.4/Sub.2/36, 2003.

105 Ibid., paragraphs 62 and 67.
106 Ibid., paragraph 4; S. Pridan-Franck, 'Human-Genomics: A Challenge to the Rules of the Game of International Law', *Columbia Journal of Transnational Law* 40, 2002.
107 Commission on Human Rights Resolution 2003/83, working paper prepared by Sub-Commission Expert Ms. Florizelle O'Connor.
108 For example, Opinion 54 of The French National Consultative Ethics Committee in 1997, where it cited Article 16–14 of the Civil Code which prohibits genetic modification of descendants as an implicit ban on reproductive cloning. Another example is the Law on Healthcare in Georgia, which states 'Human cloning through the use of genetic engineering methods shall be prohibited'. In the Sixth Committee, a delegate was of the view that 'the great expansion in knowledge of the human genome could lead to important discoveries … and also to serious abuses (referring to cloning)' (A/C.6/56/SR.27 at paragraph 11). See also C. Kuppuswamy, *et al.*, *Is Human Reproductive Cloning Inevitable?: Future Options for UN Governance*, Tokyo: UN University-Institute of Advanced Studies, 2007, p. 8.
109 Convention on the Grant of European Patents (EPC), opened for signature 5 October 1973, 13 ILM 268 (entered into force 7 October 1977).
110 Directive 98/44/EC of the European Parliament and of the Council of 6 July 1998 on the legal protection of biotechnological inventions.
111 1869 *United Nations Treaty Series* 299, reproduced in *International Legal Materials* 33, 1994, 1197.
112 The text contained in the Resolution is as follows: 'since the implementation of the TRIPS Agreement does not adequately reflect the fundamental nature and indivisibility of all human rights, including the right of everyone to enjoy the benefits of scientific progress and its applications, the right to health, the right to food, and the right to self-determination, there are apparent conflicts between the intellectual property rights regime embodied in the TRIPS Agreement, on the one hand, and international human rights law, on the other hand.'
113 'Human Rights and Bioethics', expanded working paper submitted by Ms. Iulia-Antoanella Motoc in accordance with Sub-Commission decision 2002/114, UN Doc. E/CN.4/Sub.2/36, 2003, at paragraph 45.
114 ECOSOC Doc. E/C.12/1998/SR.49, 1998, paragraph 62.
115 D. Schroeder, 'Human Rights and Their Role in Global Bioethic', *Cambridge Quarterly of Healthcare Ethics* 14 (2), 2005, 221–3.
116 M. Hartlev, 'Biotechnology – A Challenge to Human Rights?', paper given at the 9th World Congress of Bioethics, Rijeka, Croatia, 6 September 2008; K. Turković, 'Informed Consent: Facing Challenges of Multiculturalism', plenary talk at the 9th World Congress of Bioethics, Rijeka, Croatia, 6 September 2008.
117 ECOSOC Doc. E/CN.4/Sub.2/2004/38, 2004.
118 ECOSOC Doc. E/CN.4/Sub.2/2005/38, 2005.
119 ECOSOC Doc. E/CN.4/Sub.2/2004/38, 2004, paragraphs 15 and 16.
120 The United Nations Declaration on the Rights of Indigenous Peoples A/RES/61/295 was adopted in 2007 by a majority of 144 states in favour, four votes against (Australia, Canada, New Zealand and the United States) and 11 abstentions (Azerbaijan, Bangladesh, Bhutan, Burundi, Colombia, Georgia, Kenya, Nigeria, Russian Federation, Samoa and Ukraine).
121 ECOSOC Doc. E/CN.4/Sub.2/2005/38, 2005.
122 ECOSOC Doc. E/CN.4/Sub.2/2005/38, 2005, paragraph 4.
123 ECOSOC Doc. E/CN.4/Sub.2/2005/38, 2005, paragraph 9.
124 United Nations Development Programme, *Making Global Trade Work for People*, pp. 221–2.
125 P. M. Gerhart, 'Why law making for global intellectual property is unbalanced', *European Intellectual Property Review* 22 (7), 2000, 309–13, 309.

126 Ibid.
127 'Global strategy and plan of action on public health, innovation and intellectual property', WHA Res. 61.21, World Health Assembly, 61st Assembly, Agenda Item 11.6, 24 May 2008.
128 UN Doc. E/CN.4/Sub.2/36, 2003, paragraph 9.
129 WHO Framework Convention on Tobacco Control, WHA Res. 56.1, World Health Assembly, 56th Assembly, 4th Plenary Meeting, Agenda Item 13, Annex, WHO Doc. A56.VR/4, 21 May 21 2003. Available online at http://www.who.int/tobacco/fctc/text/en/fctc_en.pdf; see also K. Shibuya, *et al.*, 'WHO Framework Convention on Tobacco Control: development of an evidence based global public health treaty', *British Medical Journal* 327, 2003, 154–7.
130 *International Health Regulations*, 2nd edn, Geneva: WHO, 2005. Under the WHO constitution powers are granted to the Specialised Agency to adopt legally binding measures for the purpose of preventing the international spread of disease, which Member States can opt out from if they wish to (Articles 21(a) and 22). Currently 198 states are subject to IHR 2005.
131 D. P. Fidler, *International Law*, Geneva: WHO, available online at http://www.who.int/trade/distance_learning/gpgh/en/index.html (accessed 2 February 2008).
132 Advisory Committee on Health Research, *Genomics and World Health*, Geneva: WHO, 2002.
133 WHA Res. 56.27, World Health Assembly, 56th Assembly, Agenda Item 14.9, 28 May 2003.
134 The first official amendment of the WTO takes into account this concern and it now enables production of pharmaceutical products under compulsory license for export to 'eligible' countries (TRIPS, Article 31*bis*). This amendment has not yet come into force.
135 Article 2, WHA Res. 56.27, World Health Assembly, 56th Assembly, Agenda Item 14.9, 28 May 2003.
136 See the last section of Chapter 1.
137 United Nations Millennium Declaration, General Assembly Resolution, 55th Session, UN Doc. A/RES/55/2, 18 September 2000.

3 The common heritage of mankind (CHM) in international law

3.1 The concept of the CHM

The CHM concept is a philosophical concept, a legal principle, and in the context of the human genome, a biological concept too. It has been discussed by leading jurists as both a philosophical and legal concept. Four hundred years ago, when Hugo Grotius, the 'father of international law', wrote on the concept, he declared that the seas are the CHM, and that everybody should be able to use them.[1] This paved the way for the establishment of global trade routes and led to the opening up of cultures and countries to each other. Grotius was thus able to keep the sea lanes in other parts of the world open for ships from his own country. A modern day analogy would be the internet revolution caused by a technology that is openly accessible with the facility to share ideas and develop collaboratively through the formation of a network where the digital lanes are always open.[2]

During the mid-1960s newly independent countries, almost all of them with very poor economic growth and wealth, were trying to find strategies to achieve economic and social development. Natural resources were their main source of income and when the possibility of mining such resources from the seabed arose with the development of deep seabed mining technology and the discoveries of vast reserves of metals and minerals on the seafloor, these countries sought to neutralise any reduction of value in their land resources by the potential abundance under the sea. It was argued that in order to prevent unfair exploitation of seabed minerals, an equitable framework of sharing was necessary. The CHM concept served as the platform to push their concerns regarding underdeveloped economies forward and has been instrumental in bringing to the forefront the 'inequities' in global trade policy, especially in relation to raw materials. The concept had the support of both developing and developed countries and can be construed, at one level, as a response to the urge of humankind to reach out to others and form the UN in order to uphold the interests of humanity, where the dignity and worth of the human person is recognised first and foremost and where the common interest of humanity can be served.

As a philosophical concept, its content is vague[3] and open to myriad interpretations.[4] However, some fundamental elements of a CHM regime can be identified based on its use in international treaties. This book adopts the interpretation of an earlier work on the concept.[5] The following are the key elements of a treaty-based CHM regime:[6]

- Non-appropriation
- International management
- Benefit sharing
- Peaceful use.

When Ambassador Arvid Pardo of Malta introduced the concept of the CHM in the General Assembly in 1967[7] he insisted on the right of everybody to use it, both formally and actually.[8] The CHM concept was not about preservation and non-use – it was about active exploitation in order to benefit equitably. In elaborating the Maltese view, he stated that the CHM could not be appropriated, so that it would be kept open to use by the international community.[9] He advocated an understanding of the CHM which required the international community to bring full and comprehensive benefits to users – not just financial ones but 'also benefits derived from shared management and transfer of technology'.[10] Although his vision was broader than the finally agreed text on common heritage in the law of the sea, especially with respect of transfer of technology,[11]the regime supports a framework which provides for the sharing of resources by all stakeholders, which is the whole of mankind. In its usage by Grotius the emphasis on fair use was not obvious, although it can be inferred that he thought it was only fair that the Dutch ships should be allowed free passage through the sea lanes. In the 1830s it was again used by Andres Bello to mean that things held in common patrimony are those things 'that could not be held by one nation without detriment to the others'. In modern discussions of the term CHM, since the mid-twentieth century, the occurrence of the phrase 'fair and equitable' has increased in the General Assembly and treaty negotiations. In a later chapter in this book, it will be demonstrated that this is justified and necessitated by the rule of law, based on a fundamental framework of human rights.

In international law the CHM concept was applied to various resources during the postwar decades of the twentieth century. The most recent negotiations involved plant genetic resources (PGR). The international community was occupied with the governance of PGR before they addressed the issue of the human genome. The following section will trace the international legal regime applicable for PGR and analyse its significance to the debate on human genome governance and the principle of the common heritage of humanity.

3.2 The international legal framework for plant genetic resources (PGR)

The first ever international agreement on access to PGR important for food and agriculture was adopted at the 1983 Conference of the Food and Agriculture Organization (FAO).[12] This was before the adoption of the Convention on Biological Diversity (CBD) and over a hundred countries adhered to the International

Undertaking on Plant Genetic Resources for Food and Agriculture. A series of agreed interpretations of the agreements were adopted at subsequent FAO conferences.[13] Article 1 of the 1983 International Undertaking states that 'plant genetic resources are a heritage of mankind and consequently should be available without restriction'.[14] This was supplemented by the FAO resolution which also stated that 'plant genetic resources are a heritage of mankind to be preserved, and to be freely available for use, for the benefit of present and future generations.'[15]

The agreement was subject to changes following the adoption of various resolutions at the FAO. Sovereign interests in plant genetic resources and the interests of farmers and plant breeders were elaborated in these resolutions. In 1991 States moved to change the CHM clause in the International Undertaking. The concept of the CHM, interpreted as free availability of PGR to all those that wanted to use them, was restricted and subjected to sovereign rights. The resolution at the FAO states 'the concept of mankind's heritage, as applied in the International Undertaking on Plant Genetic Resources, is subject to the sovereignty of the states over their plant genetic resources'. In 1992 the CBD was signed and it granted sovereign rights to States in the matter of biodiversity, which includes PGR. This was accompanied by the adoption of a Resolution by States participating in the conference to look into the 'Interrelationship Between the Convention on Biological Diversity and the Promotion of Sustainable Agriculture'[16] in order to harmonise provisions between the two international documents. The Resolution urges States to find 'ways and means [...] to develop complementarity and co-operation between the Convention on Biological Diversity and the Global System for the Conservation and Sustainable Use of Plant Genetic Resources for Food and Sustainable Agriculture'.[17]

In response to this, the FAO passed a resolution in 1993 to begin negotiations to revise the International Undertaking. The CHM principle was replaced by sovereign rights and the concept of common concern for biodiversity was instead applied to PGR. This created controversy in the international community[18] and States took another eight years to agree on a revised version of the International Undertaking, which was adopted in 2001 as the International Treaty on Plant Genetic Resources.[19] The concept of CHM was dropped and the Convention recognised sovereign rights of States over their PGR. The International Treaty also envisages that 'states may mutually benefit from the creation of an effective multilateral system for facilitated access to a negotiated selection of these resources and for the fair and equitable sharing of the benefits arising from their use'.[20] It now has 111 States Parties, entered into force on 29 June 2004 and is a hybrid of the 1983 International Undertaking and the 1992 CBD.[21]

3.3 The impact of the PGR/CHM debate on the governance of human genome

The lengthy debate on access and use rules for PGR culminated in the establishment of sovereignty over PGR. The CHM principle was replaced by the principle of sovereign rights over PGR. Around the same time, in 1994, the Secretary-General of the UN was campaigning to get States to sign the Convention on the Law of the Sea. The resulting Implementation Agreement[22] contained modifications to the 1982 Convention in order to make the provisions of the Convention more acceptable to certain developed States like the US which raised objections, especially to the deep seabed mining regime of the Convention based on the CHM principle. On the whole critics of the CHM principle were pointing to the demise of the principle, and claimed that it no longer could play a useful role in the governance of resources.[23]

But in the case of debates on governance of the human genome, UNESCO established the International Bioethics Committee (IBC) in 1993 and one of the first principles enunciated in the forum was that of the common heritage of humanity.[24] The reasons for this are manifold. The shared nature of the human genome is one of the main reasons the principle of the CHM was found to be especially suitable for the regulation of activities relating to the human genome. The issue of equal contribution of all members of the human species to the shaping of our genetic heritage[25] featured high on the agenda of regulating this new emerging area of science, with allusions being made to the fruit of labour theory in intellectual property (IP) law. The human genome was, in itself, considered the typical heritage of humanity. It cannot be subject to territorial limits, nor is it possible to address it in isolation from human beings, for whom it is of instrumental value. It is shared by all members of the human species who also contribute to the shaping of it. Despite the fact that there are differences in the make-up of individual human genomes, it is an entity whose nature has been collectively determined over centuries, through generations of human evolution. Therefore it is literally the common heritage of humanity.

Unlike PGR, the human genome is intimately tied to the protection of human rights. The CHM principle, with its focus on equitable sharing of resources, was considered to be a suitable principle on which a benefit-sharing regime could be developed.[26]

Despite setbacks to the CHM principle during the early 1990s, it still continues to occupy a place within the policy framework for the human genome and its relevance to benefit-sharing regimes in human genome research continues to be explored. The CHM debates in the PGR regime have provided valuable discourse on the notion of CHM in modern times and serve to highlight the complex issues involved in regulation of genetic resources.

This chapter attempts to analyse the various components of a CHM regime and establish that the 1997 Universal Declaration on Human Rights and the Human Genome contains more of a CHM regime than is apparent from a first reading. Article 1 of the Universal Declaration states: 'The human genome underlies the fundamental unity of all members of the human family, as well as the recognition of their inherent dignity and diversity. In a symbolic sense, it is the heritage of humanity.' Although Article 1 qualifies the common heritage of humanity with the word 'symbolic', the body of the Universal Declaration is underpinned by the provisions which make up a CHM regime. The proposed CHM regime is further strengthened by the principles elaborated in the 2005 Universal Declaration on

Human Rights and Bioethics, which is the second notable instrument adopted by UNESCO enumerating general principles. The first section of this chapter analyses the 1982 UN Convention on the Law of the Sea and identifies the elements of a CHM regime therein. The second section identifies a CHM regime in the 1979 Agreement Governing the Activities of States on the Moon and Other Celestial Bodies (more commonly known as the Moon Treaty). Section 3 examines the 1997 Universal Declaration and outlines the CHM regime embedded within its provisions. The final section revaluates the 1997 Universal Declaration as a human rights declaration and identifies second generation human rights. This analysis also underlines the proposition that the CHM regime for the human genome is closely aligned to human rights principles and norms. This section provides the pathway through which the concept is reinterpreted as a human rights-centric concept in the next chapter.

3.4 The UN Convention on the Law of the Sea: deep seabed minerals

'Deep-seabed minerals' is the term commonly used to denote the substances found beneath or on the surface of the ocean floor. The most commonly studied deep seabed minerals are manganese nodules, also known as polymetallic nodules. They were discovered by a British research expedition during the nineteenth century. Manganese nodules are defined by the Mining Code as 'one of the resources of the Area consisting of any deposit or accretion of nodules, on or just below the surface of the deep seabed, which contain manganese, nickel, cobalt and copper'.[27] The recently discovered polymetallic sulphides, cobalt-bearing crusts, methane hydrates, oil, gas, marine phosphorites and deposits of precious metals also fall within the category of deep seabed minerals. The 'Area' refers to the seabed area beyond the limits of national jurisdiction, which extends beyond the outer limit of the continental shelf.[28] According to Article 136 of the 1982 UN Convention on the Law of the Sea,[29] 'The Area and its resources are the common heritage of mankind'. The CHM doctrine in the Convention on the Law of the Sea is a fundamental principle that underlies the basic principles on seabed mining. It binds all parties to the convention; there can be no reservation against the operation of this Article by any Member State and the scope for amendment of this provision is limited. The coming into force of this UN Convention and the Implementation Agreement of July 1994,[30, 31] have effectively put an end to the freedom of mining the high seas and installed a CHM regime in its place.[32] The deep seabed minerals that are found within 200 miles of the baselines are not governed by the CHM regime and are subject to the national jurisdiction of the respective States. The law of the sea related to deep seabed mining has undergone significant changes due to the introduction of the concept of the CHM. In 2001, the whole system of deep seabed mining, based on the CHM concept, became operational when the International Seabed Authority (ISA) executed contracts with pioneer investors from China, India, Japan, Korea, Russia and others.[33] It creates obligations on States to share expertise in scientific research, protect and preserve the environment, share mining technology, deal fairly in terms of prices, keep the market well and truly open, prevent monopolies in trading, provide economic assistance to States whose economies rely on land-based minerals, keep the ocean floor open for peaceful purposes, and take into consideration the interests of all States, whether landlocked or coastal.

The UN Convention has reversed the application of the long-standing freedom of high seas mining principle and replaced it with a system where there is international cooperative exploitation of resources, with due consideration for the environment. Even those countries which have not ratified the Convention and have in place unilateral mining rules, such as the US,[34] have accepted that the CHM concept underpins the governance of the oceans, although with reservations about the operation some of its fundamental principles.

3.5 The elements of the CHM regime

The elements of the CHM can be found in Part XI of the 1994 Convention and should be read in conjunction with Annexes III and IV concerning conditions of prospecting, exploration and exploitation and the administrative mechanism to go with it.[35] The Implementation Agreement in July 1994[36] has modified the mechanism.

3.5.1 Non-appropriation

Article 137 facilitates the equitable distribution of benefits from the ocean floor and the seabed by stipulating against claims of an exclusive nature by countries that might be able to stake such claims through their technological prowess and financial resources. This provision of the Convention prevents discovery and occupation of the seabed by unilateral action. In the absence of the non-appropriation principle, property rights become attached to the seabed, followed by the various benefits. It states 'No State shall claim or exercise sovereignty or sovereign rights over any part of the Area or its resources, nor shall any State or natural or juridical person appropriate any part thereof. No such claim or exercise of sovereignty or sovereign rights nor such appropriation shall be recognised'. Subsection 2 of Article 137 reiterates that 'all rights in the resources of the Area are vested in mankind as a whole, on whose behalf the Authority shall act.' It goes on to state that 'No State or natural or juridical person shall claim, acquire or exercise rights with respect to the minerals recovered from the Area except in accordance with this Part.'[37] While the non-appropriation principle applies absolutely to the seabed or the ocean floor, the minerals recovered from it are not bound by the same principle.

3.5.2 International management

The institutional framework for the CHM regime provides for proportional participation based on equity considerations for various State parties to the Convention. Countries are grouped together into five groups on the basis of mineral consumption, investment capacity, export capacity (including at least two developing countries), their status as developing countries (with special interests, including landlocked States, island States and major mineral importer States) and their geographical location (Africa, Asia, Eastern Europe, Latin America and the Caribbean and Western Europe and others). Eighteen members have been chosen based on the fifth criteria, six members have been chosen from among the developing countries and four members from the first three groups. Such a system ensures that deep seabed mining is managed collectively by States and all decisions related to the regime are taken at the international level. The long period of time over which negotiations for the Convention on the Law of the Sea took place have facilitated detailed assessments of the various economic and geographical characteristics of countries.

The negotiations to amend the Convention to make it acceptable to certain developed countries involved further prolonged discussions, and led to compromises and concessions on the part of the developing countries, but the Area is still internationally managed and special consideration is given to those countries whose economies are underdeveloped in order that they benefit from deep seabed resources. The Pioneer Investor Scheme was one such later innovation, and was grafted onto the mining regime in order to take account of the mining activities and investment already made by groups of countries, which included both developing and developed countries, and this arrangement was agreed upon by the international community.

The ISA is a Convention body and holds centralised control over all mining activities in the seabed Area. It is composed of three different principal organs and three committees: the Enterprise, which is in charge of mining activities, has a separate legal personality and can exercise independent judgement on a number of substantial issues relating to mining;[38] the Council; and the Assembly. The latter two take all the important decisions relating to the functioning of the ISA. Decision making is by consensus in all organs of the ISA and if consensus cannot be reached the decision is arrived at through a two-thirds majority vote. In case the Assembly does not agree with the Council on any matter, it shall return the matter to the Council for further consideration.[39] The Assembly elects the members of the Council according to the rules provided in Section 3(15) of the Implementation Agreement. The modified rules for representation in the Council were agreed upon after intense negotiations leading up to the 1994 Implementation Agreement and were further outlined in the ISA's discussions in 1996.[40] The countries whose high consumption rates of minerals gave them the leverage in the negotiations and countries whose investments would be indispensible in the exploitation of the resources obtained a greater voice at the table. However, since the principle of CHM had already been established and accepted, it was necessary for them to make compromises to include other interests. Representation is the first step in a system of international management. A number of 'Special Interests' were accommodated – States which were landlocked, island States, the least-developed countries, countries that were major importers of minerals of the category found in the seabed and States with large populations. Besides this, equitable geographical representation was another criteria necessitated by the notion of international management. Overall, the system of mining under the CHM regime has been welcomed as a successful international drafting effort and one that would bring about uniformity and stability in international ocean governance, balancing the interests of 'industry, the environment, consumer states, land-based producer states, developed states and developing states.'[41]

3.5.3 Benefit sharing

According to Article 140 of the Convention, the role of the ISA is to 'provide for the equitable sharing of financial and other economic benefits derived from activities in the Area through any appropriate mechanism, on a non-discriminatory basis'. It also stipulates that 'the benefit of mankind as a whole, irrespective of the geographical location of States, whether coastal or land-locked' should be taken into account, and that 'particular consideration of the interests and needs of developing States and of peoples who have not attained full independence or other self-governing status' should be given in the course of equitable sharing.[42] This article is further strengthened by Article 148 which deals with the participation of developing States in activities in the Area and requires that mining activities should be carried out with 'due regard to their special interests and needs, and in particular to the special needs of the land-locked and geographically disadvantaged among them'.[43] Article 150 touches on the practical aspect of the benefit for mankind – the policies relating to mining. It states 'Activities in the Area shall, as specifically provided for in this Part, be carried out in such a manner as to foster healthy development of the world economy and balanced growth of international trade, and to promote international cooperation for the over-all development of all countries, especially developing States'. Article 151 deals with production policies and stipulates that:

> the Assembly shall establish a system of compensation or take other measures of economic adjustment assistance including cooperation with specialized agencies and other international organizations to assist developing countries which suffer serious adverse effects on their export earnings or economies resulting from a reduction in the price of an affected mineral or in the volume of exports of that mineral, to the extent that such reduction is caused by activities in the Area.

The reason for this provision was that land-based mineral producing countries would be hard hit by competition from sea-based minerals resulting in collapse of the economy of States which rely heavily on raw material export for their earnings.[44]

Transfer of technology was one of the key areas for the effectiveness of the CHM concept in the convention. The provisions for the technology transfer in the convention were modified by the 1994 Implementation Agreement in order to reduce the obligations on technologically developed countries in sharing such technology and to make it attractive for developed countries to become parties to the Convention.[45] The modified provisions on technology transfer do not provide for full disclosure of technology but enable joint ventures between developing and developed countries. Technology transfer is now governed by Article 144, in conjunction with section 5, paragraph 1 of the Annex of the Implementation Agreement. The provisions in the Implementation Agreement state that those countries seeking to obtain technology shall do so 'on fair and reasonable commercial terms and conditions on the open market, or through joint-venture arrangements'. It also states that such acquisition must be 'consistent with the effective protection of intellectual property rights'.[46] The convention also encourages international technical and scientific cooperation and training programmes in these areas for countries that do not have the expertise. The scaling down of technology transfer provisions resulted mainly from the Implementation Agreement. Protection of IP was on the top of the agenda in 1994, and during the run-up to the formation of the World Trade Organization (WTO) and the TRIPS Agreement. It would have undermined the effort to globalise and harmonise IP protection if the technology transfer provisions had been left unchanged.

Article 143 commits States to conducting marine scientific research for the benefit of mankind as a whole. Article 242 promotes international cooperation in scientific research and Article 254 encourages States to take into consideration the needs of landlocked countries. Protection of the marine environment is an important objective within the CHM regime in the law of the sea. Article 145 of the Convention stipulates that the Authority should adopt rules for the prevention of pollution, protection and preservation of the marine flora and fauna.

3.5.4 Peaceful use

Article 141 of the Convention categorically rules out non-peaceful uses of the seabed. It states 'The Area shall be open to use exclusively for peaceful purposes by all States, whether coastal or land-locked'. Peaceful use of the seabed was an important part of the debates on the CHM because of the very real threat to disruption of international peace and security from nuclear weapons that could be placed in the seabed and ocean floor.[47] Treaty negotiations lasted through the period of the Cold War, and the danger of nuclear war was not unreal. As a parallel, such a provision would not be out of place in human genome governance frameworks in the light of the threat of bio-terrorism.

3.5.5 Operation of the regime

The system of equitable exploitation of deep seabed resources in the Area is governed by the above-mentioned main principles. In practice, it is based on the basic regime found in Annex III of the Convention, the Mining Code[48] developed by the ISA in 2000 and the Pioneer Investor Scheme. It is a complex system and one that is not yet being put to test. The ISA carries out the administration of the CHM regime. The dispute-settlement mechanism has compulsory jurisdiction over certain types of disputes. The system of mining under the CHM regime incorporates many innovative provisions designed to bring about equitable results in deep seabed mineral exploitation. Prospecting, exploration and exploitation are the three stages in mining, as recognised in the convention.

Prospecting for polymetallic nodules in a certain specified area within the Area can be conducted by prospectors with the permission of the ISA, for an unlimited period of time. Prospecting does not confer any title on the minerals in the area. Exploration and exploitation can be carried out by the Enterprise, which is the executive arm of the ISA[49] and by States Parties or juridical persons on behalf of States Parties. Exploration can be carried out after the ISA has approved a plan of work submitted by the parties.[50] Applications by parties other than the Enterprise should include plans conforming to the site-banking system. This is a system whereby 'the applicant should give the Authority the opportunity to reserve one half of the proposed contract area for exploitation through the Enterprise or in association with developing states'. The 'one half' then becomes the reserved area and is based on the mineral content and commercial value rather than on the actual dimensions. The ISA banks the reserved area for its future use. The plans of work are essentially contracts between the ISA and the mining operators.

Considerations of equity play a further role in the anti-monopoly clauses that are part of the CHM regime. Any plan of work which covers an area exceeding 'in size 30 per cent of a circular area of 400,000 square kilometres surrounding the centre of either part of the area covered by the proposed plan of work' would not be approved by the ISA.[51] Any individual State is also limited to exploring and exploiting less than 2 per cent of the total seabed area 'which is not reserved or disapproved for exploitation pursuant to Article 162, paragraph 2(x)'.[52] The contract for exploration is limited to a period of 15 years and power is vested with the ISA to grant extensions for a period of five years at a time. The ISA also controls the production of minerals. The production policy is based on equity and fairness and broadly covers unfair economic practices[53] and the provision of economic assistance[54] to developing States. An economic assistance fund established by the ISA was conceived of in order to provide assistance to developing land-based mineral producers on determination that they have been hard hit by sea-based mineral production. The ISA is authorised to conduct studies to determine which economies are affected and to what extent they have been hard done by. International trade rules apply to production policies and disputes are to be settled through relevant multilateral trade agreement rules and through the Convention's dispute settlement authority. The financial independence of the ISA is secured in the Convention through the provision for three types of payments: the application fee, the annual fixed fee, and the royalty system. The application fee was fixed at 250,000 USD for the approval of a plan of work for one phase.[55]

An annual fixed fee has to be paid to the ISA once commercial production is commenced. The implementation agreement provides for a royalty system or a combination of a royalty and profit-sharing system.[56] While the earlier Convention's provisions provided for a more detailed and complicated system of benefit sharing, the Implementation Agreement rejected those provisions and left it open-ended for rules to be formulated at a later time by the ISA. The currency of payments was flexible and not confined to the major currencies in the world economy and there was the option that payments to the ISA could be made in the equivalents of processed metals at market values.[57] There is an obligation on the contractors[58] to provide training for personnel of the ISA and developing States, with a focus on the conduct of exploration.[59]

The system of exploitation of deep seabed resources is unique, one that is not found elsewhere for any other resource. It is closely regulated by sovereign powers working cooperatively through a system agreed upon by consent. It allows for production and exploitation of resources, and although it is regulated at the stage of production, its later use is left completely to market forces. Given that other resources and indeed minerals compete with it in the global market, it will not completely work to the detriment of technology-rich countries by creating an equalising effect in the global economy, but it would give a much needed boost to the low-income economies.

3.6 The Moon Treaty

If the CHM regime in the Convention on the Law of the Sea can be described as ambitious, in terms of laying out an equitable regime for a resource which was not yet technologically accessible on a large scale, the Moon Treaty is even more so, as it is not yet clear what type of resources are on the moon, let alone the uses they might be put to. It was finalised in 1979 and came into force on 11 July 1984. Article 11 of the Moon Treaty proclaims 'The moon and its natural resources are the common heritage of mankind'. Paragraph 5 of Article 11 commits States to elaborating a CHM regime for the exploration and exploitation of the moon. Such use should be based on the principles of non-discrimination and equality and in accordance with other provisions of international law.[60] The treaty provides for review conferences in order to keep track of technological advancements in exploration and space activities.[61] State parties also have an obligation to inform the UN Secretary-General, the public and the international community of any resources that they may discover on the moon.[62] The emphasis in the CHM principle in the Moon Treaty is on the use and the expansion of opportunities in the use of moon resources.[63] The treaty becomes especially relevant after the new plan unveiled by the US administration to explore the moon and its resources,[64] and in the light of developments in space exploration by Asian countries.[65]

3.6.1 Non-appropriation

According to Article 11(2):

> The moon is not subject to national appropriation by any claim of sovereignty, by means of use or occupation, or by any other means. Neither the surface nor the subsurface of the moon, nor any part thereof or natural resources in place, shall become property of any State, international intergovernmental or non-governmental organisation, national organisation or non-governmental

> entity or of any natural person. The placement of personnel, space vehicles, equipment, facilities, stations and installations on or below the surface of the moon, including structures connected with its surface or subsurface, shall not create a right of ownership over the surface or the subsurface of the moon or any areas thereof.

Article 6(2) states that:

> In carrying out scientific investigations and in furtherance of the provisions of this Agreement, the States Parties shall have the right to collect and remove from the moon samples of its mineral and other substances. Such samples shall remain at the disposal of those States Parties which caused them to be collected and may be used by them for scientific purposes. States Parties shall have regard to the desirability of making a portion of such samples available to other interested States Parties and the international scientific community for scientific investigation. States Parties may in the course of scientific investigations also use mineral and other substances of the moon in quantities appropriate for the support of their missions.

The surface of the moon and the subsurface of the moon are not subject to unilateral exercising of property rights under this regime. The regime seeks to prevent a race for the resources on the moon in order to either occupy it or to exploit its resources.

3.6.2 International management

The Moon Treaty requires states to develop an international regime for the orderly and safe development of the natural resources of the moon and its rational management.[66] The Treaty recognises the freedom of scientific research.

3.6.3 Benefit sharing

Article 4 of the Moon Treaty lays down the basic principle of benefit sharing of moon resources and states that exploration 'shall be carried out for the benefit and in the interests of all countries, irrespective of their degree of economic or scientific development.' It facilitates the taking into consideration of the disparities in 'economic and social progress' and states the need for exploration to be beneficial in promoting higher standards of living conditions'. The Article stipulates that international cooperation should involve both mutual assistance on a multilateral or bilateral basis and international organisations. An equitable sharing by all States Parties in the benefits derived from those resources, whereby the interests and needs of the developing countries, as well as the efforts of those countries which have contributed either directly or indirectly to the exploration of the moon, shall be given special consideration.[67]

3.6.4 Peaceful use

According to Article 3 of the Moon Treaty, 'The moon shall be used by all States Parties exclusively for peaceful purposes. No use of force is allowed on the moon, nor should it be used as a base for use of force on earth or man-made objects in space.' Any military use of the moon, except for scientific research for peaceful purposes, is prohibited.[68]

The Moon Treaty and the Outer Space Treaty[69] provide a comprehensive regime for resource management on the moon and other celestial bodies.

In the last few decades, States have come together to regulate the use of resources which have been either newly discovered or are considered to have the potential to become strategically and economically useful by declaring them the CHM. This enables States to collectively develop rules to regulate their exploitation and to achieve peaceful coexistence of peoples. As international trade has become more globalised, States have shifted focus and become engaged in ensuring that trade and commerce is free and fair, underpinned by the notion that free trade is the vehicle for prosperity, and that international peace and security could best be achieved by promoting trade agreements. This conviction continues to prevail as the underlying reason behind the growing importance of the WTO and such bodies as the World Intellectual Property Organization (WIPO). But since the end of the twentieth century, health and environment issues have exposed the need for regulation of international trade and free market-oriented arrangements in order to achieve health and environment goals and ensure peace and security. The human genome, as a special and valuable resource, could be governed collectively under a system of equitable exploitation or be subject to numerous national and international regulations. The 1997 Universal Declaration attempted to develop one such model of the CHM governance of the human genome based on previous treaty models. It is a matter of debate whether the model outlined in the Universal Declaration is suitable, but the international community should promote such a debate. It would contribute to the strengthening of the international rule of law by leading to the development of coherent fundamental principles for various natural resources deemed to be of international significance.

3.7 The CHM and the human genome

The earliest references to genetic heritage are evident in the scoping meetings of the Scientific and Technical Orientation Group which met in 1993 to identify the issues that were to be brought into the purview of the body drafting an international instrument to protect the human genome.[70] The human rights commission also considered the issues intersecting bioethics and human rights and in a resolution passed by the UN Commission on Human Rights, States were urged to report to the Secretary-General any activities that were being carried out to ensure that life sciences develop in a manner 'beneficial to humanity as a whole'[71].

The first solid appearance of the CHM was in a report submitted by the Director-General of UNESCO to the General Conference on the 'revised outline of a declaration on the protection of the human genome'[72], which considered the use of a CHM regime for the genome. The report suggested that the concept of the CHM, as found in international law, should be understood in the context of the human genome as the idea of 'equitable sharing among countries of the advantages it is hoped that its [human genome's] exploitation may provide'[73]. It further refers to the objective of fair exploitation that is embedded in the notion of the CHM and guarantees access to the results of research on the human genome to 'all people and of all countries, especially the developing countries'.[74] Speaking at the second meeting of the IBC, which had been established to develop international governance regimes for the human genome, Justice Bedjaoui, the then president of the International Court of Justice (ICJ) and a member of the IBC, explored the notion of the CHM and its applicability to the human genome.

According to Justice Bedjaoui, 'any advance in knowledge about the human genome must benefit humankind as a whole'. He also stated that a CHM regime is mindful of 'the inequalities in the development of the various regions of the world', adding that 'the scientific data collected on the human genome must be used to eradicate certain diseases affecting particular areas of the world, or factors responsible for the underdevelopment of the human body in particular climates or environments'.[75] Justice Kirby, another highly respected member of the IBC and a member of the legal commission which drafted the international instrument, interpreted the CHM concept to mean that 'the benefits of advances should be made available to all.' He cautioned against the situation where individuals from developing countries would be the subjects of such research through the use of their tissue samples. To counter this he suggested that human solidarity be a core element of the 1997 Universal Declaration, and amongst other things, that developing countries benefit from, and are not merely the object of, the research.'[76]

The IBC discussed the issue of the governance of genome research in the context of benefit sharing as well as in the context of the protection of the rights of the individual human subject and the researcher.[77] It pioneered the attempt to bring together different types of human rights such as the right to development (although it does not receive direct mention), the right to privacy, the right to protection of one's data, the freedom of research, the right to informed consent, the right against discrimination, the right to life, etc. The innovative feature of the Universal Declaration was the attempt to define the human genome as the CHM and to provide it with a CHM framework of governance. The attempt met with some success. The various elements of a CHM regime found a place in the Universal Declaration and the issues of distributive justice and equity in genome research were reasonably addressed. And since it was only the first step towards the establishment of a governance regime for the human genome, it is a good start.

Upon the adoption of the 1997 Universal Declaration at the General Conference, State representatives from Japan and Germany stated that they viewed it to be the beginning of the engagement of the international community on rules governing the human genome. The Japanese representative stated that his government did not regard the Universal Declaration to be an 'unchangeable instrument' but as a reflection on the human genome and human rights.[78] The German representative regarded the declaration 'as the beginning of a global dialogue and the development of protective standards in the field of research on the human genome and its application'.[79] The Observer from the US endorsed the Universal Declaration and stated that it was an 'important step forward in providing a framework for inquiry on the human genome'.[80] The representative from Argentina was categorical about the continuing nature of the debate on governance of the human genome when he stated that the Universal Declaration does not conclude the debate on this subject; it was, he held, 'a point of departure for future clarification of matters which still need to be defined'. He referred to the changes that are constantly taking place in the field of science and technology as one of the reasons for keeping the debate going.[81] Representatives from Canada, Israel and Brazil did not explicitly state that the Universal Declaration was the start of a process of consideration of ethical issues relating to the human genome but expressed their interest in participating in the *ad hoc* working group for the follow-up to the Universal Declaration. The General Assembly, by endorsing the Universal Declaration[82] the following year, has firmly established the issues relating to the human genome on the international agenda and it has continued to occupy centre stage in ethics discussions at the international level.[83]

3.8 The CHM elements in the 1997 Universal Declaration

Article 1 of the 1997 Universal Declaration states that the human genome is 'symbolically the heritage of humanity'. From the above analysis of the discussions among the Drafting Committee, it is possible to conclude that the notion of 'the heritage of humanity' corresponds to the notion of the CHM found in international law. The December 1996 draft outline of the Universal Declaration contained the provision 'the human genome is the common heritage of humanity'.[84] The final version of the draft document, dated 20 December 1996, states 'The human genome is common heritage of humanity. It underlies the fundamental unity of all members of the human family, as well as the recognition of the inherent dignity of each of its members.' The following analysis identifies the CHM elements in the Universal Declaration.

3.8.1 Non-appropriation

Article 4 of the 1997 Universal Declaration is widely held to be a non-appropriation clause.[85] It states 'The human genome in its natural state shall not give rise to financial gains.' The genome is a unique natural resource. It is naturally present and functional in all human beings. Information held in one person's genome is very similar to that of the information held in another's and this is true of the whole human population. This non-appropriation clause has to be interpreted based on the unique nature of this natural resource. The non-appropriation clause in the seabed regime applies to the seafloor and not directly to the minerals themselves – which give the seabed its value – and is the justification for making the land under water not subject to exclusive rights. Drawing parallels between the seabed and the human genome in its natural state is therefore not implausible.

Intellectual property rights (IPR) protection can be granted to any part of the genome. The Universal Declaration's non-reference to IPR does not preclude the granting of such protection. As one author puts it, 'the status of the Human Genome as a *res omnium*[86] *[Hugo Grotius' term for the CHM]* would not, per se, bar private appropriation of parts thereof'.[87] At the General Conference, Canada said of the non-appropriation clause that there was a 'need to clarify, in relation to existing intellectual property norms, and that the human genome in its natural state refers to the total genome in a human body'.

The Observer from the US was more specific in his statement and said that the US understood the text to mean 'that the human genome in its natural state *is a discovery and not an invention* and thus shall not give rise directly to financial gain.'[88] The intention here is to recast the terminology in the language of patent systems – an invention is eligible for a patent but a discovery is not. The Brazilian representative was more inclined to leave the interpretation open and stated that the Universal Declaration 'does not address with precision, the issue of appropriation of the Genome'. This however was not the intention of the drafters, based on the explanatory note accompanying the Universal Declaration, which clearly states that 'simple knowledge of human genes, or partial sequences of genes, in their natural state should not be directly used to secure financial gains'. However, the human genome itself is not defined anywhere in the Universal Declaration[89], nor is there an accepted definition of what the human genome is, meaning there is some room for ambiguous understanding based on the elaboration given in the explanatory note.

In conclusion, it can be stated that the CHM regime in the Universal Declaration applies in the non-appropriation clause to the genome in its natural state only.

3.8.2 International governance

The Member States of UNESCO welcomed the work of the IBC and endorsed the role that it played in framing rules for the governance of the human genome. No direct mechanism for international governance is specified in the 1997 Universal Declaration; Article 19 purports a framework for international cooperation with the participation of relevant international organisations supporting initiatives prompted by States. The IBC was entrusted with an important role to 'further examine issues raised by the application of the principles in the declaration and by the evolution of the technologies in question through appropriate consultations with parties concerned, such as vulnerable groups.' It was also entrusted the task of making 'recommendations, in accordance with UNESCO's statutory procedures, addressed to the General Conference and to give advice concerning the follow-up of this Declaration, in particular regarding the identification of practices that could be contrary to human dignity, such as germ-line interventions'.[90] Seemingly, a mechanism of collective governance finds favour in the Universal Declaration. It is useful to recall the nature of the human genome and the nature of human experimentation, which in essence transcends boundaries and is not amenable to effective national governance. The need for an international regime to monitor and lay down rules for the governance of human reproductive cloning and the subsequent attempt to deliver an international treaty on the subject are also indicators of the need for international governance of the human genome.[91] A probable 'International Research Ethics Committee' would be able to take into consideration the interests of humanity and balance risks at a global level. Article 21 of the Universal Declaration provides guidance for a process by which cosmopolitan interests could be taken into account, by 'an open international discussion, ensuring the free expression of various sociocultural, religious and philosophical opinions.' Such a body could undertake the delivery of benefit-sharing arrangements.

3.8.3 Benefit sharing

The benefit-sharing provisions in the 1997 Universal Declaration occupy considerable space. They are based on the idea that progress in genomic science and technology would bring about benefits that should be shared equitably between populations. The type of benefits are neither defined nor mentioned. From the Preamble, which states that 'research on the human genome and the resulting applications open up vast prospects for progress in improving the health of individuals and of humankind as a whole', the nature of benefits can be construed as health benefits. The Universal Declaration recognises the importance of the freedom of research, as a good in itself as well as being essential for benefit sharing. It promotes freedom of research and limits it to no more than 'meticulousness, caution, intellectual honesty and integrity' in carrying out research and in the presentation and utilization of findings.[92] The onus is on the science policy makers to clarify these responsibilities.[93] States are responsible for ensuring that they facilitate the freedom of research, subject only to 'safeguarding respect for human rights, fundamental freedoms and human dignity and the protection of public health'.[94]

Article 12 of the Universal Declaration deals with health. It stipulates that the 'applications of research, including applications in biology, genetics and medicine, concerning the human genome, shall seek to offer relief from suffering and improve the health of individuals and humankind as a whole'.[95] The Universal Declaration makes a link between 'biology, genetics and medicine'[96] and proposes that benefits 'arising from advances in these branches of knowledge shall be made available to all'.[97]

Article 15 supports the protection of public health. The provisions under Section E enunciate a policy which primarily benefits public and global health. Entitled 'Solidarity and International Co-operation', this section contains three articles and urges States to promote solidarity amongst families, communities and different population groups. It also entreats States to espouse a principle of solidarity beyond borders and recommends they 'should foster, inter alia, research on the identification, prevention and treatment of genetically based and genetically influenced diseases, in particular rare as well as endemic diseases which affect large numbers of the world's population.'[98] Article 18 lays the onus on States 'to continue fostering the international dissemination of scientific knowledge concerning the human genome, human diversity and genetic research and, in that regard, to foster scientific and cultural co-operation, particularly between industrialised and developing countries.' Article 19 proposes specific criteria for cooperation:

> a In the framework of international co-operation with developing countries, states should seek to encourage measures enabling:
>
> i assessment of the risks and benefits pertaining to research on the human genome to be carried out and abuse to be prevented;
> ii the capacity of developing countries to carry out research on human biology and genetics, taking into consideration their specific problems, to be developed and strengthened;
> iii developing countries to benefit from the achievements of scientific and technological research so that their use in favour of economic and social progress can be to the benefit of all;
> iv the free exchange of scientific knowledge and information in the areas of biology, genetics and medicine to be promoted.
>
> b Relevant international organizations should support and promote the initiatives taken by states for the above-mentioned purposes.

The benefit-sharing provisions are extensive and potentially open-ended. Such loosely worded provisions might seem beneficial to developing states but will not favour developed States. But Section E does not stand alone. Although there is potential to interpret solidarity and international cooperation provisions in the Universal Declaration in the broadest sense, this would not be in keeping with the fundamental concept underpinning the Universal Declaration – human dignity.

Respect for, and protection of, human dignity is the underlying value of the Universal Declaration.[99] It is integral to the notion of benefit sharing as much as it is an overarching prerogative throughout the instrument. Article 2 gives expression to the value by affirming that:

> a Everyone has a right to respect for their dignity and for their rights regardless of their genetic characteristics.
> b That dignity makes it imperative not to reduce individuals to their genetic characteristics and to respect their uniqueness and diversity.

Dignity is used and simultaneously abused in bioethics.[100] According to Beyleveld and Brownsword, 'it is an abuse of the concept of human dignity to operate it as a veto on any practice that is intuitively disliked'.[101] Article 2's conception of dignity is based on respect for individual rights irrespective of their genetic characteristics and does not abuse the term dignity since it provides for an anti-discriminatory clause, which is well documented within international human rights law and also within the Universal Declaration itself (Article 6). However, dignity in the Universal Declaration can be taken to mean more than a reaffirmation of negative rights. Health and solidarity provisions shape its notion of dignity. It aims to promote genomic research whose goal is to 'relieve suffering and improve the health of individuals and humankind as a whole'.[102] Human dignity incorporates both positive and negative rights. It is also not an open-ended concept. Section E of the Universal Declaration is a human rights-based promotion of solidarity between families and populations supported by States via collaborative research and the sharing of technology. It strives to achieve human dignity.

The Member States of the World Health Organization (WHO) adopted a resolution acknowledging the 'remarkable progress made in genomics research' and adopted a number of activities covering benefit sharing in order to benefit less developed countries. It recommends States 'find creative and equitable ways of mobilising more resources for genomics research targeted at the health needs of developing countries'.[103] This resolution further strengthens health as an essential issue in genome research.

As an aside, it is worth noting that the Universal Declaration does not go as far as it might have done on the subject of human dignity. Article 11 simply states 'Practices which are contrary to human dignity, such as reproductive cloning of human beings shall not be permitted'. A more appropriate reference to the restoration of dignity in the context of concerns such as global health emergencies[104] or other global priorities[105] would have been more helpful.

Benefit sharing is further enhanced by Article 21 of the Universal Declaration which urges States to actively promote capacity building and dissemination of human genomic research.

3.8.4 Peaceful use

Article 15 of the 1997 Universal Declaration covers the freedom of research fundamental to the progress of society and recommends that States ensure that 'research results are not used for non-peaceful purposes.' The tangible threat to peace and security arising from genetic technology can be in the form of spreading disease through genetically modified organisms in war or in times of peace. The threat from genetically enhanced beings to non-enhanced beings is also to be foreseen.

3.9 The Legal Commission debates: elements of a CHM regime or isolated principles?

The various elements identified above were indeed conceptualised as part of a CHM regime, and not as isolated principles. Speaking at the IBC plenary sessions one of the members stated:

> We are now proposing to include the human genome in the common heritage of humanity. Legally speaking, this would be a historic and revolutionary measure, fraught with implications and attended by many consequences that would be beneficial to humanity.[106]

While the plenary discussions on the CHM bring out the direction and thinking behind the notion as applied to the human genome,[107] the Legal Commission also added its own discussions on the concept, much along the same lines as the plenary sessions, and strengthened the notion of CHM in the Universal Declaration. The members of the Legal Commission mainly consisted of academic lawyers working in the areas of public and private international law and members of State judiciary and the ICJ. The Legal Commission decided to work towards drafting a declaration on human rights and bioethics, based on the reasoning that a non-binding declaration would then lead to a binding agreement, as has been the case in many international human rights instruments.[108] Another model proposed was that of private international law, which features the adoption of a single-article convention along with an annex of the general principles on the issue. These could then be followed up by supervisory committees established to monitor the application of the draft model law.[109] The choice was however made to draft a declaration on human rights and the human genome, along with the rejoinder that a convention would be good but was not feasible at that point due to constraints on time. It was expected that the IBC might decide to have something at hand to commemorate the 50th anniversary of the Universal Declaration of Human Rights.[110,111] The terminology of the title of the instrument, i.e. 'the Protection of the Human Genome', was contentious. It was criticised on the basis that the first and foremost concern should be the protection of human rights and not the protection of the human genome.

It was contended that 'the protection of the human genome should also, and above all, be based on human rights'.[112] The members of the legal commission were in no doubt that the operative section on the CHM 'formed the cornerstone of innovation introduced by the declaration'.[113] Although the biological notion and concerns in private law were raised in the course of discussions on the CHM, 'the general feeling was that it was more appropriate to refer to the broader concept of the "common heritage of humanity" enshrined in international law.'[114] The draft declaration was sent out to a 100 institutions and specialists to elicit their opinion on its provisions.

In the Fourth Session of the Legal Commission, the Chairman dwelt at length on the notion of the CHM in his speech to the members of the commission. Tracing the history of the concept from the early nineteenth century as enunciated by Andres Bello and Paul de Geouffre de la Pradelle he noted that the legal principles based on the CHM in the law of the sea, the moon and celestial bodies and cultural objects, thereby further entrenching the international law concept of the CHM into the Declaration.[115] In an effort to stimulate discussion on the concept so that the implications of a CHM regime could be fully understood and assessed, the International Federation of Philosophical Societies went on record stating that the 'declaration that the human genome is the common heritage of humanity was not fully understandable'.[116] The scope of the CHM notion was explored in the general discussion on the concept. On the one hand, 'bringing the notion of common heritage into play in the declaration was an attempt to secure the broadest possible protection of the human genome against damage that might endanger the very survival of humankind'.[117] On the other hand, however, 'the notion of heritage covered the knowledge accumulated by men and women about themselves as a source of potential for the progress of humankind'.[118]

In the fifth meeting of the Legal Commission, the initial results of the consultation on the 'outline of a declaration on the protection of the human genome' were discussed. The President opined that the objections that were raised regarding the qualification of the human genome as the heritage of humanity 'stemmed perhaps from an inadequate understanding of this legal concept'.[119] The President stressed that the term 'common heritage of humanity' should be maintained in the declaration 'in view of the need to guarantee respect for human dignity and human rights and the need for a balance between the protection of individual rights and the common interest of humanity'.[120] With regard to the issue of the patentability of gene sequences, the President of the IBC stated that 'the concept of the common heritage of humanity was not incompatible with the prospect of patentability of the human genome'.[121] She added that it would be difficult to address the issue of patentability 'within the context of a general document such as the declaration on the human genome', and also that such an 'important and complex issue' (given the economic stakes) cannot be addressed by UNESCO alone.[122] The results of the international consultation exercise led to changes in the draft declaration and a revised version of the preliminary draft declaration was discussed at the Sixth Session of the Legal Commission. There was a strong move towards amending the title of the instrument to refer to human rights and not to the protection of the human genome. The consultation exercise further skewed the equation by bracketing patenting and the CHM concept together and the significance of the CHM as an equitable framework for resource allocation and the torch bearer of human rights and human dignity was dimmed. It all came down to just one issue: 'Did describing the human genome as the common heritage of humanity imply prohibiting the patenting of human genes?'[123]

At the Sixth Session, during a discussion on individual articles in the declaration, it was agreed that Article 1 would 'simply stipulate that the human genome was a fundamental component of the common heritage of humanity'.[124] At the Seventh Session of the Legal Commission, the President of the ICJ acknowledged the revolutionary nature of the CHM concept in a declaration on the human genome and human rights and held that the notion was not being evoked in a 'furtive' way. He suggested that more of the Declaration should reflect the fundamental nature of the concept that underpins it.[125] He underscored the importance of the CHM notion by stating that 'it could be affirmed if, apart from the preamble, the text consisted of only two articles. Article 1 would proclaim the human genome as common heritage of humanity, while Article 2 would stipulate the adoption of the Declaration by the states, the remainder of the operative part stating the principles to be complied with would be placed as an annex'.[126] In a general discussion on the CHM concept and Article 1 of the Declaration, several members felt that it was an indispensable part of the instrument, while others felt that 'the notion of a common heritage could create tension between the individual heritage of a person and the heritage of the species, each gene is thus deemed to be the property of the individual and an inalienable property of humanity'.[127] During the eighth meeting of the Legal Commission, the final version of the draft declaration, which was to be forwarded to the IBC, was discussed. Section A of the draft declaration contained 'the common heritage of humanity'. This article was considered the cornerstone of the future declaration.[128]

Far from being a set of rules formulated to govern human genome research, the debates at the legal commission reveal that this was an orchestrated attempt to bring together a CHM regime for human genome governance. As Justice Bedjaoui put it:

> the human genome can perfectly well be regarded as a 'heritage' as it possesses the specific property of any heritage, which is that it is transmitted. Everyone, from conception, inherits this heritage, which develops, dwindles, changes and is transmitted with them. The human genome is then the heritage, the *hereditas*, of all human beings.[129]

The international community should now consider building a binding CHM regime for the human genome.[130]

At the Fifth Session of the IBC in 1998, the Director-General of UNESCO referred to the adoption of the 1997 Universal Declaration on the Human Genome and Human Rights by the General Conference as a landmark and historic event.[131] He applauded the Member States for having transcended religious, cultural and political differences to proclaim the human genome the heritage of humanity.[132]

3.10 Human rights and the human genome

The 1997 Universal Declaration contains a CHM framework and also relies on a human rights framework. In its Preamble, it refers to the International Bill of Rights, the core group of international law human rights instruments,[133] and a number of humanitarian law and International Labour Organization documents. It mentions UNESCO resolutions on the call for progress in science and technology to conform to human rights standards,[134] which was the subject of a General Assembly resolution in 1975.[135] The last paragraph of the Universal Declaration states that"

> research on the human genome and the resulting applications open up vast prospects for progress in improving the health of individuals and of humankind as a whole, but … such research should fully respect human dignity, freedom and human rights, as well as the prohibition of all forms of discrimination based on genetic characteristics.

The human rights aspects of genome governance were further strengthened and deepened by the 2005 Universal Declaration on Human Rights and Bioethics. This has a wider scope than the 1997 Universal Declaration and contains a further set of human rights. The right to health is elaborated in greater detail in the 2005 Universal Declaration. It is a step towards addressing some of the doctrinal issues involved in human rights and bioethics.[136] The need for fundamental clarification of the relationship between human rights and bioethics has been addressed in the

2005 Universal Declaration. Article 14 dwells on the right to health and enunciates a broad view of health including access to quality health care and essential medicines, access to adequate nutrition and water, improvement of living conditions and environment, elimination of the marginalisation and exclusion of persons on any grounds, and the reduction of poverty and illiteracy.[137]

Article 15 of the 2005 Universal Declaration, entitled 'Sharing of Benefits', specifies particular methods of benefit sharing such as access to quality health care,[138] provision of new diagnostic and therapeutic modalities or products stemming from research,[139] support for health services,[140] access to scientific and technological knowledge,[141] and capacity-building facilities for research purposes.[142] The 2005 Universal Declaration urges particular emphasis should be laid on developing countries.[143] Despite criticism from bioethicists about the principles contained in this instrument[144] it is still laudable that the debate engaged many more stakeholders at the international level.

Both the 1997 Universal Declaration and the 2005 Universal Declaration elaborated by UNESCO rely on a strong framework of second generation human rights or, in other words, economic, social and cultural rights (ESCR).[145] Human genome governance can be satisfactorily achieved only if there is a satisfactory system of recognition and justification of second generation rights. The notion that human rights are indivisible[146] is common knowledge to human rights lawyers,[147] but there is a hierarchy in the way the rights are fulfilled. The right to health and the right to development are victims of neglect.[148] It is as expensive and resource intensive to run a good court system and police force as it is to put in place a good health infrastructure and spend resources on ensuring that the right to health is a reality for everyone.[149] Yet most countries have a better resourced policing and criminal justice system than health system. Doctrinally it is a convincing argument that the distinction between positive and negative rights are false[150] yet in practice negative rights have been recognised and upheld, mainly because of the 'ease' of enforcement with respect to these rights.[151] Most of the well-known human rights advocacy organisations monitor civil and political rights[152] but not ESCR.

An issue that affects the enforcement of second generation human rights is the determination of the extent of the protection granted. According to one view, every right has a core and a penumbra, a maximum and a minimum. So, under the 'maximin' definition of the right to health,[153] the States have a duty to, at the very minimum, protect individuals against serious health threats and, at the maximum, they should strive to fulfil the attainment of the highest possible standard of health for all individuals.[154] If such a conception of a core (minimum) and penumbra (maximum) conception of ESCR is possible[155], it is also possible to reconceptualise civil and political rights in a similar manner. Why is there a maximum conception of civil and political rights? If all rights are equal and indivisible, then why not offer a 'maximum-minimum' rights protection system to civil and political rights too? It would increase the possibility that the core ESCR could be fulfilled.

3.11 Conclusion

The experience with the governance of PGR did not prevent the IBC from proposing a CHM framework for human genome governance. The 1997 Universal Declaration has sketched out a CHM framework that warrants further debate. A CHM framework, developed from previous models available in international law, does not prohibit the operation of IPR in genomic research. If this is the main concern with the CHM concept, it is unfounded. The 1997 Universal Declaration is also a human rights framework and one that is reliant on a strong system of ESCR, which in themselves are not robust enough. To pitch for a right to health and its fulfilment (Article 14, 2005 Universal Declaration) is a bold attempt to provide support to the right to health. A human rights approach to the governance of human genomic research would be piecemeal and ineffective so long as fundamental issues about ESCR, the so-called second generation of human rights, are not addressed. A rational theory of human rights, to effectively help fix ESCR, will be provided in Chapter 5 and will underpin an examination of the ideas of right to property, health and development.

Notes

1 Grotius' *Mare Liberum* was commissioned by the Dutch East India Company to defend their actions in the Straits of Malacca in 1603. Grotius wrote a legal defence statement based on his thesis that a fundamental principle of the law of nations included '*Every nation is free to travel to every other nation, and to trade with it'. He used the notion of res omnium to formulate the principle of the freedom of the high seas, which has ever since been applicable. It is also a part of the present day law of the sea, although subject to more restrictions than in Grotius' time.* J. B. Scott (ed.) and R. v. D. Magoffin (trans.), *Hugo Grotius, The Freedom of the Seas*, Oxford: Oxford University Press, 1916.

2 L. Lessig, *The Future of Ideas: The Fate of the Commons in a Connected World*, New York: Random House, 2001.

3 See Brazil's statement on the concept at the 24th Session of the General Assembly, 1969, in *DOA Law of the Sea* at 113. An excerpt of it is reproduced here: 'If mankind had always restricted itself to applying the legal concepts that already existed, legal systems would not have developed and law would not have fulfilled its proper function in social progress.'

4 S. Gorove, 'The Concept of Common Heritage of Mankind: A Political, Moral or Legal Innovation?', *San Diego Law Review* 9, 1972, 390–403.

5 K. Baslar, *The Concept of Common Heritage of Mankind in International Law*, The Hague/Boston/London: Martinus Nijhoff Publishers, 1998, p. 82.

6 Declaration of Principles Governing the Seabed and the Ocean Floor and the Subsoil Thereof, Beyond the Limits of National Jurisdiction, UN Doc A/RES/25/2749, 1970.

7 Agenda Item: Examination of the question of the reservation exclusively for peaceful purposes of the sea-bed and the ocean floor, and the subsoil thereof, underlying the high seas beyond the limits of present national jurisdiction, and the use of their resources in the interests of mankind, UN Doc. A/C.1/PV.1516, 1967. Link to speech available online at http://www.un.org/depts/los/convention_agreements/convention_historical_perspective.htm (accessed 28 July 2006).

8 A. Pardo, 'The Origins of the 1967 Malta Initiative', *International Insights* 9 (2), 2003, 65–9, available online at http://website.lineone.net/~romweb/par_a01.htm (accessed 28 June 2006).

9 A. Pardo, 'Law of the Sea Conference – What Went Wrong', in Freidheim (ed.), *Managing Ocean Resources: A Primer*, pp. 137–9, cited in L. F. E. Goldie, 'A Note on Some Diverse Meanings of the Common Heritage of Mankind', *Syracuse Journal of International Law and Commerce* 10, 1983, 69–112 at 87.
10 Ibid.
11 Section 5(b), Agreement relating to the Implementation of Part XI of the United Nations Convention on the Law of the Sea of 10 December 1982, 28 July 1994 (Implementation Agreement), *Law of the Sea Bulletin* 1994: Special Issue IV (Misc. No. 44) at 10. The modified provisions on technology transfer do not go as far as possible in sharing but enable joint ventures between developing and developed countries. Technology transfer is now governed by Article 144, in conjunction with Section 5, paragraph 1 of the Annex of the Implementation Agreement. The provisions in the Implementation Agreement state that those countries seeking to obtain technology shall do so 'on fair and reasonable commercial terms and conditions on the open market, or through joint-venture arrangements'.
12 International Undertaking on Plant Genetic Resources for Food and Agriculture, FAO Doc. RES/22/8, 1983.
13 FAO Doc. RES/25/4, 1989; FAO Doc. RES/25/5, 1989; FAO Doc. RES/26/3, 1991.
14 Annex, FAO Doc. RES/22/8, 1983.
15 Ibid.
16 Resolution 3 of the Nairobi Final Act, reproduced in *International Legal Materials* 1992: 31, 842–7 at 846.
17 Ibid., Article 2.
18 N. D. Hamilton, 'Who Owns Dinner: Evolving Legal Mechanisms for Ownership of Plant Genetic Resources', *Tulsa Law Journal* 28, 1993, 587–657 at 601.
19 FAO Doc. RES/31/3, 2001.
20 Ibid., Preamble.
21 G. Rose, 'International Law of Sustainable Agriculture in the 21st Century: The International Treaty on Plant Genetic Resources for Food and Agriculture', *Georgetown International Environmental Law Review* 15, 2003, 583–632 at 631.
22 Agreement relating to the Implementation of Part XI of the United Nations Convention on the Law of the Sea of 10 December 1982, 28 July 1994 (Implementation Agreement), *Law of the Sea Bulletin* 1994: Special Issue IV (Misc. No. 44).
23 D. S. Tilford, 'Saving the Blueprints: The International Legal Regime for Plant Resources', *Case Western Reserve Journal of International Law* 30, 1998, 373–446 at 415.
24 Study Submitted by the Director-General Concerning the Possibility of Drawing up an International Instrument for the Protection of the Human Genome, UNESCO Doc. C/RES/27/45, 1993.
25 D. M. Gitter, 'Resolving the Open Source Paradox in Biotechnology: A Proposal for a Revised Open Source Policy for Publicly Funded Genomic Databases', *Houston Law Review* 43, 2007, 1475–1521.
26 S. Pridan-Frank, 'Human-Genomics: A Challenge to the Rules of the Game of International Law', *Columbia Journal of Transnational Law* 40, 2002, 619–74.
27 As defined in UNCLOS III's Conference Resolution II, para. 1 (d) and Reg 1 (3)(d) of the mining code.
28 Article 1(1), UN Convention on the Law of the Sea, 1982 defines Area as 'the sea-bed and ocean-floor and subsoil thereof, beyond the limits of national jurisdiction'.
29 The UN Convention on the Law of the Sea, reproduced in *International Legal Materials* 1982, 21, 1261; UN Doc. A/ CONF.62/122, opened for signature 10 December 1982.
30 Agreement relating to the Implementation of Part XI of the United Nations Convention on the Law of the Sea of 10 December 1982, 28 July 1994 (Implementation Agreement), *Law of the Sea Bulletin* 1994: Special Issue IV (Misc. No. 44) at 10.
31 The UN Convention on the Law of the Sea came into force on 16 November 1994, 12

months after the date of deposit of the 60th instrument of ratification, in accordance with Article 308. The Implementation Agreement adopted in 1994 substantially amended the provisions on deep seabed mining and embodied a compromise, in order to persuade the industrialised countries to sign up to the convention. According to Article 2(1) of the Implementation Agreement, the Convention provisions on deep seabed mining and the changes to the provisions found in the Implementation Agreement were to be considered as one single instrument. Therefore the deep seabed mining provisions are spread over two separate international instruments. Together they came into force on 28 July 1996, when the conditions laid down in Article 3 of the Implementation Agreement had been fulfilled. Provisional Application of the Implementation Agreement and Interim Budgetary Arrangements under Article 7 also applied during the period between the coming into force of the Convention and the coming into force of the Implementation Agreement.

32 E. D. Brown, *Sea-bed Energy and Minerals: The International Legal Regime, Vol.2: Sea-Bed Mining*, The Hague: Kluwer Law International, 2001, Chapter 2.

33 J. C. Nelson, 'The Contemporary Seabed Mining Regime: A Critical Analysis Of The Mining Regulations Promulgated By The International Seabed Authority', *Colorado Journal of International Environmental Law and Policy*, 16, 2005, 27–75 at 37.

34 There is a strong indication that the US will soon join the multilateral Convention of the Law of the Sea. See D. B. Sandalow, 'Law of the Convention: Should the U. S. join?', *Policy Brief 137, The Brookings Institution*, 2004, available online at http://www.brook.edu/comm/policybriefs/pb137.htm (accessed 18 August 2006) and *Law of the Sea (2-2-05)*, American Geological Institute, available online at http://www.agiweb.org/gap/legis109/lawofthesea.html (accessed 18 August 2006).

35 Annex III: Basic conditions of prospecting, exploration and exploitation; Annex IV: Statute of the enterprise.

36 Agreement relating to the Implementation of Part XI of the United Nations Convention on the Law of the Sea of 10 December 1982, 28 July 1994 (Implementation Agreement), *Law of the Sea Bulletin* 1994: Special Issue IV (Misc. No. 44) at 10.

37 Article 137(3), Part XI, The UN Convention on the Law of the Sea, reproduced in *International Legal Materials* 1982, 21, 1261.

38 Agreement relating to the Implementation of Part XI of the United Nations Convention on the Law of the Sea of 10 December 1982, 28 July 1994 (Implementation Agreement), Law of the Sea Bulletin 1994: Special Issue IV (Misc. No. 44) at 10, Annex, Section 2.

39 Ibid., Annex, Section 3(4).

40 Composition of the First Council of the International Seabed Authority, UN Doc ISBA/A/L.8, 1996 and Corr.1 (2 April 1996).

41 J. I. Charney, 'Entry into Force of the 1982 Convention on the Law of the Sea', *Virginia Journal of International Law* 35, 1995, 381–404 at 394.

42 Article 140(1), Part XI, The UN Convention on the Law of the Sea, reproduced in *International Legal Materials* 1982, 21, 1261.

43 Also see Article 152(2), under Production Policies in Part XI, UN Convention on the Law of the Sea, reproduced in *International Legal Materials* 1982, 21, 1261.

44 Statement by the Vice-Chairman of Subcommittee I on (Item 2) of the Work Programme: Status, scope, functions and powers of the international machinery at the 55th Meeting held on 4 August 1972 (A/AC.138/SC.I/L.17), in *DOA Law of the Sea* at 235.

45 H. Miskell, 'Law of The Sea: The Convention Enters into Force', *American Society of International Law Proceedings*, 89, 1995, 451–70 at 451.

46 Section 5(b), Agreement relating to the Implementation of Part XI of the United Nations Convention on the Law of the Sea of 10 December 1982, 28 July 1994 (Implementation Agreement), *Law of the Sea Bulletin* 1994: Special Issue IV (Misc. No. 44) at 10.

47 See debates in the Committee on the Peaceful Uses of the Seabed and the Ocean Floor beyond the Limits of National Jurisdiction, in *DOA Law of the Sea*.

48 Regulations on Prospecting and Exploration for Polymetallic Nodules in the Area, UN Doc., ISBA/6/A/18, as approved by the Assembly on 13 July 2000.
49 Article 158(2), The UN Convention on the Law of the Sea, reproduced in *International Legal Materials* 1982, 21, 1261; E. D. Brown, *Sea-bed Energy and Minerals: The International Legal Regime, Vol.2: Sea-Bed Mining*, The Hague: Kluwer Law International, 2001, p. 318.
50 Article 153(3), UN Convention on the Law of the Sea, reproduced in *International Legal Materials* 1982, 21, 1261.
51 Article 6(3)(c)(i), Annex III, UN Convention on the Law of the Sea, reproduced in *International Legal Materials* 1982, 21, 1261.
52 Article 6(3)(c), Annex III, UN Convention on the Law of the Sea, reproduced in *International Legal Materials* 1982, 21, 1261.
53 Article 151(8), UN Convention on the Law of the Sea, reproduced in *International Legal Materials* 1982, 21, 1261 and Section 6, Annex, Agreement relating to the Implementation of Part XI of the United Nations Convention on the Law of the Sea of 10 December 1982, 28 July 1994 (Implementation Agreement), Law of the Sea Bulletin 1994: Special Issue IV (Misc. No. 44) at 10.
54 Article 151(10), The UN Convention on the Law of the Sea, *International Legal Materials* 1982, 21, 1261 and Section 6, Annex, Agreement relating to the Implementation of Part XI of the United Nations Convention on the Law of the Sea of 10 December 1982, 28 July 1994 (Implementation Agreement), *Law of the Sea Bulletin* 1994: Special Issue IV (Misc. No. 44) at 10.
55 Agreement relating to the Implementation of Part XI of the United Nations Convention on the Law of the Sea of 10 December 1982, 28 July 1994 (Implementation Agreement), *Law of the Sea Bulletin* 1994: Special Issue IV (Misc. No. 44) at 10; Annex, Section 8, paragraph 3.
56 Ibid., Annex, Section 8, paragraph 1(c).
57 Annexx III, Article 13, paragraph 12, UN Convention on the Law of the Sea, reproduced in *International Legal Materials* 1982: 21, 1261.
58 Under the Mining Code, Regulation 27(1).
59 Article 144(2) and Annex III, Article 15 and Agreement Relating to the Implementation of Part XI of the United Nations Convention on the Law of the Sea of 10 December 1982, 28 July 1994 (Implementation Agreement), *Law of the Sea Bulletin* 1994: Special Issue IV (Misc. No. 44) at 10, Annex, Section 5, paragraph 1(c).
60 Article 11(4), Agreement Governing the Activities of States on the Moon and Other Celestial Bodies, reproduced in *International Legal Materials*, 1979, 18, 1434.
61 Article 18, Agreement Governing the Activities of States on the Moon and Other Celestial Bodies, reproduced in *International Legal Materials*, 1979, 18, 1434.
62 Article 11(6), Agreement Governing the Activities of States on the Moon and Other Celestial Bodies, reproduced in *International Legal Materials*, 1979, 18, 1434.
63 Article 11(7)(c), Agreement Governing the Activities of States on the Moon and Other Celestial Bodies, reproduced in *International Legal Materials*, 1979, 18, 1434.
64 Office of the Press Secretary, The White House, 'A Renewed Spirit of Discovery: President Bush Announces New Vision for Space Exploration Program', 14 January 2004, available online at http://www.whitehouse.gov/news/releases/2004/01/print/20040114–11.html (accessed 18 August 2006).
65 G. V. Jean, 'Space Programs Poised for Major Expansion in Asia', *National Defense Magazine*, November 2008, available online at http://www.nationaldefensemagazine.org/archive/2008/November/Pages/SpaceProgramsPoisedforMajorExpansioninAsia.aspx (accessed 23 October 2008).
66 Article 11(7)(a)(b), Agreement Governing the Activities of States on the Moon and Other Celestial Bodies, reproduced in *International Legal Materials*, 1979, 18, 1434.
67 Article 11(7)(d) Agreement Governing the Activities of States on the Moon and Other Celestial Bodies, reproduced in *International Legal Materials*, 1979, 18, 1434.

68 See also Treaty on Principles Governing the Activities of States in the Exploration and Use of Outer Space, Including the Moon and Other Celestial Bodies (The Outer Space Treaty), reproduced in *International Legal Materials*, 1967, 6, 386. The main purpose of this treaty is to stop the spread of an arms race in outer space.
69 See also the Declaration on International Cooperation in the Exploration and Use of Outer Space for the Benefit and in the Interest of All States, Taking into Particular Account the Needs of Developing Countries, UN Doc. A/Res/51/122, 1996.
70 Study Submitted by the Director-General Concerning the Possibility of Drawing up an International Instrument for the Protection of the Human Genome, UNESCO Doc. C/RES/27/45, 1993, p. 5.
71 Commission on Human Rights RES/82,1995; see also E/CN.4/1995/74, 1994.
72 Report by the Director-General on the Possibility of Drawing up an International Instrument on the Protection of the Human Genome, UNESCO Doc. BIO/CIB-COMJUR/95, 1995. The first substantive section of the outline proclaims the human genome as the common heritage of humanity. Section 1 states 'The Human Genome is a fundamental component of the common heritage of humanity and needs to be protected in order to safeguard the integrity of the human species, as a value in itself, and the dignity of each of its members'.
73 Ibid. p. 15.
74 Ibid.
75 *Proceedings of the Third Session of the IBC*, 1995, available online at http://portal.unesco.org/shs (accessed 15 May 2006), p. 108.
76 Justice M. Kirby, *The Universal Declaration on the Human Genome – A Vital UNESCO Initiative*, 2nd International Summit of National Bioethics Advisory Commissions, Tokyo, Japan, 1998.
77 See Section B on the Rights of the Persons Concerned and Section D on the Conditions for the Exercise of Scientific Activity, Universal Declaration on the Human Genome and Human Rights, UNESCO Doc. C/RES/29/16, 1997.
78 Draft Report of Commission III, UNESCO Doc. C/RES/29/82, 1997, p. 27.
79 Ibid.
80 Ibid., p. 29.
81 Ibid., p. 28, 26.
82 UN Doc. A/RES/53/152, 1998; UN Doc. A/53/PV.85, 1998. The UN General Assembly endorsed the declaration and the Third Committee adopted it without a vote.
83 See Universal Declaration on Bioethics and Human Rights, UNESCO Doc. C/RES/33/36, 2005.
84 Preparation of a Declaration on the Human Genome: Report by the Director General, UNESCO Doc. BIO- 97/CONF.201/3,1996, Annex 2.
85 See the Explanatory Note of the Declaration.
86 *Res omnium* means things that cannot be owned by anybody but be used and enjoyed by everybody.
87 J. A. Bovenberg, 'Mining the Common Heritage of Our DNA: Lessons Learned From Grotius and Pardo', *Duke Law & Technology Review* 0008, 2006, paragraph 16.
88 Draft Report of Commission III, UNESCO Doc. C/RES/29/82, 1997, p. 27.
89 Ibid., p. 27.
90 Article 24, Universal Declaration on the Human Genome and Human Rights, UNESCO Doc. C/RES/29/16, 1997.
91 M. H. Arsanjani, 'Negotiating The UN Declaration On Human Cloning', *American Journal of International Law*, 100, 2006, 164–79.
92 Ibid., Article 13.
93 Ibid.
94 Ibid., Article 15.
95 Ibid., Article 12(b).
96 Ibid., Article 12(a).

97 Ibid.
98 Ibid., Section 17.
99 Justice M. Kirby, *The Universal Declaration on the Human Genome – A Vital UNESCO Initiative*, 2nd International Summit of National Bioethics Advisory Commissions, Tokyo, Japan, 1998. He discusses dignity as the core of the genome declaration.
100 See T. Caulfield, 'Human cloning laws, human dignity and the poverty of the policy making dialogue', *BMC Medical Ethics* 4 (3), 2003.
101 In D. Beyleveld, R. Brownsword, 'Human Dignity, Human Rights, and Human Genetics', *The Modern Law Review* 61, 1998, 661–81.
102 Ibid., Article 12(b).
103 Section 4, Genomics and World Health, WHO Doc. WHA57.13, 2004.
104 See 'WHO reports 10 million TB patients successfully treated under "DOTS" 10 years after declaring TB a Global Emergency', World Health Organization, available online at http://www.who.int/mediacentre/news/releases/2003/pr25/en/ (accessed 12 January 2005).
105 See the UN Millennium Declaration, UN Doc. A/RES/55/2, 2000 for Millennium Development Goals (MDGs). The Declaration takes up as its central challenge ensuring that globalisation is a 'positive force for all the world's people', since 'at present its benefits are very unevenly shared, while its costs are unevenly distributed'. The MDGs seek to halve the proportion of people who suffer from hunger, reduce by two-thirds the under-five mortality rate, reduce by three-quarters the maternal mortality ratio between 1990 and 2015 and by 2015 to have halted and begun to reverse the spread of HIV/AIDS, have halted and begun to reverse the incidence of malaria and other major diseases, halved the proportion of people without sustainable access to safe drinking-water and sanitation, achieved a significant improvement in the lives of at least 100 million slum dwellers and, in cooperation with pharmaceutical companies, provided access to affordable, essential drugs in developing countries.
106 *Proceedings of the Second Session of the IBC*, 1994, available online at http://portal.unesco.org/shs (accessed 15 May 2006), p. 106.
107 Refer to the previous chapter for a summary of the plenary session meetings.
108 *Report of the First Meeting of the Legal Commission of the IBC*, available online at http://www.unesco.org/ibc/en/genome/juridique/r1.html (accessed 24 November 2001), p. 2.
109 Ibid., p. 4.
110 Ibid., p. 5.
111 See also Methodology for the Preparation of an International Instrument for the Protection of the Human Genome, 1994, available online at http://www.unesco.org/ibc/en/genome/juridique/methodologie.html (accessed 24 November 2001).
112 *Report of the First Meeting of the Legal Commission of the IBC*, available online at http://www.unesco.org/ibc/en/genome/juridique/r1.html (accessed 24 November 2001), p. 5.
113 *Report of the Third Meeting of the Legal Commission of the IBC*, Paris: UNESCO, 1994, available online at http://www.unesco.org/ibc/en/genome/juridique/r1.html (accessed 24 November 2001), p. 2.
114 Ibid.
115 *Report of the Fourth Meeting of the Legal Commission of the IBC*, Paris: UNESCO, 1995, available online at http://www.unesco.org/ibc/en/genome/juridique/r1.html (accessed 24 November 2001), p. 2.
116 Ibid., p. 3.
117 Ibid., p. 4.
118 Ibid.
119 *Report of the Fifth Meeting of the Legal Commission of the IBC*, Paris: UNESCO, 1995, available online at http://www.unesco.org/ibc/en/genome/juridique/r1.html (accessed 24 November 2001), p. 1.

120 Ibid., p. 2.
121 Ibid., p. 3.
122 Ibid.
123 *Report of the Sixth Meeting of the Legal Commission of the IBC*, Paris: UNESCO, 1996, available online at http://www.unesco.org/ibc/en/genome/juridique/r1.html (accessed 24 November 2001), p. 2.
124 Ibid., p. 4.
125 *Report of the Seventh Meeting of the Legal Commission of the IBC*, Paris: UNESCO, 1996, available online at http://www.unesco.org/ibc/en/genome/juridique/r1.html (accessed 24 November 2001), p. 3.
126 Ibid.
127 Ibid., p. 4.
128 *Report of the Eighth of the Legal Commission of the IBC*, Paris: UNESCO, 1996, available online at http://www.unesco.org/ibc/en/genome/juridique/r1.html (accessed 24 November 2001), p. 3.
129 *Proceedings of the Third Session of the IBC*, 1995, available online at http://portal.unesco.org/shs (accessed 15 May 2006), p. 107.
130 Justice Kirby opined that elements from the three UNESCO Declarations could be brought together in a future binding international instrument on the human genome (Justice M. Kirby, 'UNESCO and Universal Principles In Bioethics: What's Next?', UNESCO International Bioethics Committee Twelfth Session, Tokyo, Japan, 15–17 December 2005). See also H. Nys, 'Towards an International Treaty on Human Rights and Biomedicine? Some Reflections Inspired by UNESCO's Universal Declaration on Bioethics and Human Rights', *European Journal of Health Law* 13, 2005, 5–8 and A. Asai, S. Oe, 'A Valuable Up-To-Date Compendium Of Bioethical Knowledge', *Developing World Bioethics* 5 (3), 2005, p. 218.
131 *Proceedings of the Fifth Session of the IBC*, 1998, available online at http://portal.unesco.org/shs (accessed 15 May 2006), p. 63, 64.
132 Ibid.
133 This consists of the following seven international treaties: the International Convention on the Elimination of All Forms of Racial Discrimination, the International Covenant on Civil and Political Rights, the International Covenant on Economic, Social and Cultural Rights, the Convention on the Elimination of All Forms of Discrimination against Women, the Convention against Torture and Other Cruel, Inhuman or Degrading Treatment or Punishment, the Convention on the Rights of the Child, and the International Convention on the Protection of the Rights of All Migrant Workers and Members of Their Families.
134 UNESCO Doc. C/RES/22/13.1, 1983, UNESCO Doc. C/RES/23/13.1, 1985, UNESCO Doc. C/RES/24/13.1, 1987, UNESCO Doc. C/RES/25/5.2 and 7.3, 1989, UNESCO Doc. C/RES/27/5.15, 1993 and UNESCO Doc. C/RES/28/ 0.12, 2.1 and 2.2, 1995.
135 Declaration on the Use of Scientific and Technological Progress in the Interests of Peace and for the Benefit of Mankind, UN Doc A/RES/30/3384, 1975.
136 T. A. Faunce, 'Will international human rights subsume medical ethics? Intersections in the UNESCO Universal Bioethics Declaration', *Journal of Medical Ethics* 31, 2005, 173–8 at 173.
137 Article 14 (2)(i–v), Universal Declaration on Bioethics and Human Rights, UNESCO Doc. C/RES/33/36, 2005.
138 Article 15 (a)(ii), 2005 Universal Declaration.
139 Ibid., Article 15 (a)(iii).
140 Ibid., Article 15 (a)(iv).
141 Ibid., Article 15 (a)(v).
142 Ibid., Article 15 (a)(vi).
143 Ibid., Article 15 (a).
144 W. Landman, U. Schüklenk, 'UNESCO "declares" universals on bioethics and human

rights – many unexpected universal truths unearthed by UN body', *Developing World Bioethics* 5 (3), 2005, iii–vi; J. R. Williams, 'UNESCO's Proposed Declaration On Bioethics And Human Rights – A Bland Compromise', *Developing World Bioethics*, 5 (3), 2005, 210–15. But see also R. Macklin, 'Yet Another Guideline? The Unesco Draft Declaration', *Developing World Bioethics*, 5 (3), 2005, 244–50.

145 M. Isfahan, V. Oosterveld (eds), *Giving Meaning to Economic, Social and Cultural Rights*, Philadelphia: University of Philadelphia Press, 2001; C. Tomushat, 'The Different 'Generations' of Human Rights: From Human Rights to Good Governance, in his book *Human Rights: Between Idealism and Realism*, Oxford: Oxford University Press, 2003, pp. 24–57.

146 'All human rights are universal, indivisible and interdependent and interrelated. The international community must treat human rights globally in a fair and equal manner, on the same footing, and with the same emphasis. While the significance of national and regional particularities and various historical, cultural and religious backgrounds must be borne in mind, it is the duty of States, regardless of their political, economic and cultural systems, to promote and protect all human rights and fundamental freedoms.' Vienna Declaration and Programme of Action (Part 1, paragraph 5), adopted by the World Conference on Human Rights, Vienna, 25 June 1993 (A/CONF. 157/24 (Part 1), Chap. III).

147 M. Winston, *On the Indivisibility and Interdependence of Human Rights*, Paper presented at the 20th World Congress of Philosophy, Boston, 10–15 August 1998, available online at http://www.bu.edu/wcp/Papers/Huma/HumaWins.htm (accessed 20 January, 2006); L. M. Keller, *The Indivisibility of Economic and Political Rights*, Book Review of A. Sen, *Development as Freedom*, New York: Knopf, 1999; E. Petersmann, 'On "Indivisibility" of Human Rights', *European Journal of International Law* 14 (2), 2003, 381–5.

148 On the right to health, see C. A. B. Toebes, *The Right to Health as a Human Right in International Law*, Antwerpen, Groningen, Oxford: Intersentia, Hart Publishers, 1999. On the right to development, see A. Sengupta, 'On the Theory and Practice of the Right to Development', *Human Rights Quarterly* 24 (4), 2002, 837–89.

149 M. C. R. Craven, *The International Covenant on Economic, Social, and Cultural Rights: A Perspective on Its Development*, Oxford: Clarendon Press, 1995.

150 A. Gewirth, *The Community of Rights*, Chicago: University of Chicago Press, 1996.

151 H. Hannum, 'Human Rights as Politics and Idolatry', *The American Journal of International Law*, 97 (4), 2003, 1006–9; M. Mutua, 'Theory and Reality in the International Protection of Human Rights', *The American Journal of International Law*, 95 (1), 2001, 255–7.

152 R. McCorquodale (ed.), *Human Rights (International Library of essays in law and legal theory. Second Series)*, Aldershot, England: Ashgate-Dartmouth, 2003; C. M. Chinkin, 'Human Rights in Global Politics', *The American Journal of International Law*, 95 (2), 2001, 472–7.

153 D. P. Fidler, *International Law and Infectious Diseases*, Oxford: Clarendon Press, 1999, p. 185.

154 Gostin and Lazzarini's 'maximin approach' is as follows. The maximum content is reflected in the following definition of the right to health: 'The duty of the state, within the limits of its available resources, to ensure the conditions necessary for the health of individuals and populations.' At a minimum, 'the state would have a responsibility, within the limits of its available resources, to intervene to prevent or reduce serious threats to the health of individuals or populations.' O. Gostin, Z. Lazzarini, *Human Rights and Public Health in the AIDS Pandemic*, Oxford: Oxford University Press, 1997, p. 29.

155 See the principle of progressive realisation under Article 12 of the International Covenant on Economic, Social and Cultural Rights (ICESCR).

4 The common heritage of mankind (CHM) and the right to development (RTD)

4.1 Introduction

The CHM concept has been emptied out and filled in time and again through the centuries. The normative version found in the 1982 UN Convention on the Law of the Sea and the 1979 Moon Treaty is but one version of the CHM. To flesh out the concept of the CHM for the human genome the background behind the Convention's framework would be instructive on two counts. First, similar concerns on health and the economy prevail in the twenty-first century as did during the postwar years. Second, seeing beyond the elements of the CHM into the problems and issues that motivated States to take collective action to address systemic economic changes leads us to human rights approaches. As alluded to in the last section of the previous chapter, human rights play a central role in the governance of the human genome. In order to place a human genome CHM framework on a human rights scaffold, an investigation into how the CHM is connected to human rights is useful, and this is elaborated on in this chapter. In the next chapter, the notion of the hierarchy of human rights is taken up. Such a hierarchy exists in practice, in the existence of stronger civil and political rights and weaker economic, social and cultural rights (ESCR). It will be argued that, theoretically, a hierarchal system of human rights is acceptable, but the current hierarchical system is not justifiable.

Looking beyond the CHM elements brings us to the RTD, which forms the bedrock of these elements. The RTD is defined as 'an inalienable human right by virtue of which every human person and all peoples are entitled to participate in, contribute to, and enjoy economic, social, cultural and political development, in which all human rights and fundamental freedoms can be fully realized.'[1] The RTD is a unique human right and should not be counted as one of a fast-growing compendium of human rights. It is a right that exemplifies the potential paradigm shift taking place in international human rights law. There is no longer a clear doctrinal divide between first, second and third generation human rights – they are collapsing into each other, giving rise to the RTD, which incorporates features of civil, political, economic, social and cultural rights. It is a fundamental right, essential for the enjoyment of all other rights.

The RTD and normative CHM frameworks developed in response to international concerns, particularly from developing States about the alarming lack of human and economy security in various regions in the world. These States put forward a set of proposals collectively called the New International Economic Order (NIEO). By analysing the UN Declaration on the Establishment of a New International Economic Order, the associated Charter of Economic Rights and Duties of States, and the emergence of the human rights approach to development from the ashes of the NIEO – including the intervening development of the CHM normative models as a continuum in search of non-institutional and institutionally based arrangements for economic and social justice – this chapter comes to the conclusion that the international community of States retains faith in a human rights approach to governance. The RTD is fundamental to the existing CHM normative models.

The relevance of the RTD to the governance of the human genome is apparent in the content of the right, which consists of the fulfilment of basic health needs. A comparison of the provisions of the 1986 Declaration on the Right to Development and the 1997 Universal Declaration on the Human Genome and Human Rights plots out similar concerns such as wellbeing, solidarity and international cooperation.

4.2 The New International Economic Order (NIEO)

4.2.1 Background

In the 1960s the international community was made up of a growing number of developing States, newly independent States and States struggling for independence. With the Cold War at its peak,[2] powerful interests in the world were polarised into opposing blocs with numerous lesser powers taking sides. However, a group of countries remained non-aligned and called themselves the Non- Aligned Movement (NAM). Initially focusing on military and security issues, from 1964 onwards NAM discussions included economic and political issues. The economies of developing countries were seriously affected because of trade and economic imbalances caused by various factors, both internal and external.[3] NAM was formed to facilitate space for the discussion of these issues on the global stage. At the same time, the UN Conference on Trade and Development (UNCTAD) was formed in 1964 to bring about more equitable and free trade, underpinned by concerns for the poorest economies. The Group of 77 grew out of UNCTAD and currently consists of 130 States. UNCTAD sought to protect prices for raw materials and facilitated commodities arrangements to enable stable and fair prices for raw materials from the poorest economies. Developing countries demanded price stabilisation for a range of commodities exported by developing countries, the tying up of raw material prices with the corresponding manufactured goods prices, and transfer of technology[4] – all of which would substantially increase the value of raw materials within their territories.

The dialogue on the economic system established after the Second World War was dominated by the major powers, with a few developing countries participating merely as 'showpieces',[5] their voices going unheeded. The rules of the game were also based on the needs of the postwar economies of Western Europe and America.[6]

The Bretton Woods Institutions were operating without consideration of human rights in trade and were applying models that were unsuitable for economies in different parts of the world.[7] The geopolitics at the time affected the formulation of international instruments almost in every area. Human rights law was also affected and the resulting emphasis on civil and political rights did further damage to the rights of peoples in the developing and newly emerging countries.

The UN provided a platform for countries on an equal basis, so that the newly independent and other populations could voice their ideas on issues that affected their wellbeing and prosperity. Developing countries united together in this cause and sought to exercise their collective bargaining power.[8] A radical expression of the mood of the States can be captured in the following passage:

> The historical and political reasons for the present disorder can be mainly expressed in terms of imperialism, colonialism and neo-colonialism. Dependence, exploitation, the looting of the resources of the Third World, and the introduction of zones of influence, have marked international relations with 'organised' or 'institutionalised' disorder. The cruel, inhuman law of maximum profit has finally succeeded in establishing disorder, with the Faustian power of multinational firms, gigantism of military-industrial complexes and the ecological disaster.[9]

Critics decried the movement as nothing more than a means to enable powerful individuals within developing countries to gain more wealth and power by manipulating international arrangements.[10] While this criticism was not unfounded, such a view ignored other ground realities in relation to unfair and unsuitable trade relations. The NIEO was a 'demand for more justice for the ordinary people of the developing countries'.[11]

4.2.2 The NIEO as a human rights movement

The NIEO movement took a more particular approach to the problems identified by the General Assembly in 1961.[12] The Member States pinpointed the problem of the growing gap between the per capita incomes of economically developed and less developed countries and that the rate of economic and social progress in developing countries was still far from adequate, despite efforts having been taken to address the issue.[13] The NIEO movement was a human rights movement from the very beginning because of the way it conceived of international cooperation and solidarity. It attempted to hold developed State economies responsible for the uplifting of the poorer economies. A direct cause-and-effect link was made between the poverty and underdevelopment of the peoples in developing countries and the economic policies of the developed countries, and the CHM too owes its formulation to the NIEO. The use of the term 'mankind' in the CHM concept is noteworthy as it refers to a collection of individuals and not States.

With increasing visibility through increased representation in the General Assembly, Third World countries brought about a resolution declaring the 1960s as the 'Development Decade'. The goal of the Development Decade was to:

> accelerate progress towards self-sustaining growth of the economy of the individual nations and their social advancement so as to attain in each under-developed country a substantial increase in the rate of growth, with each country setting its own target, taking as the objective a minimum annual rate of growth of aggregate income of 5 per cent at the end of the Decade.[14]

The measures that were to be undertaken during the decade were the modification of trading rules in primary commodities, intensification of international cooperation and efforts towards economic growth targets, and the review of the role of all economic players towards the achievement of the goals of the declaration. Specific mention was made of the need for accelerating measures to eliminate 'illiteracy, hunger and disease, which seriously affected the productivity of the people of the less developed countries'.[15]

The first UN Development Decade did not achieve its goals as States failed to achieve target growth rates,[16] and the second UN Development Decade was launched.[17] In the opening statement, States committed themselves to the goal of creating 'conditions of stability and well-being and to ensure a minimum standard of living consistent with human dignity through economic and social progress and development'.

During 1973 and 1974 a number of significant events took place. In 1973, oil-producing countries formed the Organization of Petroleum Producing Countries (OPEC) and the notion of producers' groups was born. The oil embargo they imposed affected the world economy and marked the beginning of OPEC as a 'permanent force to be reckoned with'.[18] Although the oil-producing countries had found their place in the international community, the other developing countries were still left behind with struggling economies. Amidst these events, Ambassador Arvid Pardo from Malta introduced the concept of the CHM as a system of deep seabed mining governance intended to address some of the issues affecting the economies of developing countries. It was an attempt to enhance the cooperation between developed and developing countries. When the Conference on the Law of the Sea commenced in 1973, the stage was set for the developing countries to bring their ideas into action. A resolution was adopted at the 6th Special Session of the UN General Assembly in 1974 on the Declaration[19] and Programme of Action[20] of the New International Economic Order. The Charter of Economic Rights and Duties of States was adopted by the General Assembly in 1974.[21] The 7th Special Session was then convened to discuss economic questions and adopted the Resolution on Development and International Economic Co-operation.[22]

4.2.3 The NIEO: its successes and failures

In rallying together with the goal of foisting radical systemic changes onto the existing rules of economic growth, Third World countries in their statements on the NIEO expressed the view that the North should also shoulder the burden for development, because their economic policies adversely affected development in the South.[23] But the First World, conceding there was some responsibility on their part to assist development, were only willing to extend help on their own terms.[24] This was deemed insufficient and perhaps even detrimental by developing countries. With such vast differences in perceptions of their responsibilities, there are bound to be both failures and successes in the various NIEO ventures. The NIEO movement was elaborate and wide-reaching; almost every UN agency and international organisation became a forum for promotion of the movement, as the States promoting it saw a fundamental role for the objectives enshrined in NIEO goals in a wide range of international activities.

The NIEO was an economic, political and a cultural movement aimed at 'challenging the status quo of the international system on various counts.'[25] Taken in this sense, it is an ongoing movement, which has managed to keep the agenda for change going, and is hence a success. Despite the claims that the 'founding' documents of the NIEO were 'immediately legally binding',[26] it is subsequent practice that has given more credence to the resolutions. The resolutions in the General Assembly are only of persuasive value and do not form hard law. Only those resolutions that have a bearing on the internal functioning of the body or those stipulated under the UN Charter[27] have binding effect. The major NIEO documents were more political in nature than legal.[28] The continuing influence of these documents has been seen in the agenda for change in international trade forums, hence it would not be fully appropriate to dismiss these documents as having no effect at all. As more policies and treaties are adopted incorporating basic NIEO principles, these declarations accrue more legal value. The principles incorporated in the NIEO declarations found support among a large group of States, even if the ways to implement such goals were very controversial.

The reinvigorated sessions at UNCTAD on commodities, development assistance, multinational corporations, transfer of technology and proposals to revise national and international patent regimes are testimony to the acceptance of the central features of the NIEO.[29] The Lomé Convention, which was adopted in 1975[30] to establish a 'new model' for relations between developing and developed countries in order to fulfil the aspiration of the international community to move towards a more just and balanced economic order, is a successful arrangement which embodies many of NIEO aspirations. It offers the opportunity for change from 'the paternalism and enlightened self interest of Europe towards African states, visible in earlier agreements'[31] and to bring about changes through the programme on stabilisation of export earnings, industrial cooperation through capacity building and limited exchange of technology, and good practices in development assistance.

The establishment of the right to self-determination and the principle of sovereignty over natural resources have been achieved through NIEO. These principles could be viewed as being part of the accepted principles in international law, but their restatement as part of the NIEO was a reminder that these principles have not always been upheld and are therefore worth reinforcing.[32] The notion of self-determination as not mere political independence and equality but as economic independence and equality was however a contribution of the NIEO movement to the discourse on self-determination. The role of multinational corporations and the erosion of sovereignty is a recurring NIEO theme,[33] which is of practical significance in the RTD, and especially in the health component of the RTD. Control over natural resources has been enshrined in international law through various instruments and it has covered new territory in recent years.[34]

The establishment of the principle of the CHM in modern international law and the attendant CHM regime of governance is in itself a revolutionary new framework of governance which can rightly be claimed to be a new international economic order for the seabed resources, the moon and outer space. In more recent years, the rise of new economies such as India, Brazil and China has been hailed as a new international economic order, and it would not be an exaggeration to state that there are links between the NIEO and the emergence of the new global powers.

Despite many advances, the movement for a new international economic order went unfulfilled. Developing States failed to impose the large-scale changes they intended to bring about. The formation of OPEC and its control over oil was in fact a new and radical usurpation of power, but did not consolidate into a wider benefit for the developing countries in general. The developing countries themselves were seen to be placing obstacles before the formation of a new international economic order. Inadequate investment, unemployment, protectionism and other restrictions, and the lack of self-reliance common in developing countries were obstacles in the development of economies.[35] By the mid-1980s, confidence in the NIEO movement was waning. In 1981, developing countries attempted to initiate a debate to further changes towards fulfilling the NIEO but an attempt to pass a resolution was later withdrawn. This was the start of the decline of the NIEO movement. In many respects the NIEO has been amalgamated with the debates on globalisation, human rights and environment protection.[36]

The UNCTAD sessions, from 1997 to date, are evidence for such a trend. The 10th Session of UNCTAD held discussions on the impact of globalisation on four key areas – finance, trade, investment and social development.[37] UNCTAD continues to stress increasing inequalities in trade and commodities.[38] The changing nature of trade and the introduction of hi-tech industries have moved the emphasis away from commodities, leaving the economies of poor countries to suffer. The movement for the NIEO, as spelt out in the founding documents, has largely lost significance in the present day and the forces that propped it up have also lost significance. Yet the problems that first prompted the movement for the NIEO continue to persist into the current century and are continuing to engage the international community.[39]

4.2.4 The Declaration on the Establishment of a New International Economic Order (1974)

The Preamble of the Declaration on the Establishment of a New International Economic Order refers to 'the purposes and principles of the Charter of the United

Nations to promote the economic advancement and social progress of all peoples'.[40] Independence from colonial dominance and technological progress feature as the main objectives of the NIEO. The existing international economic order is portrayed as detrimental to the achievement of an even and balanced development of the international community.[41]

The Declaration establishes 'the political, economic and social well-being of present and future generations' as the goal of the effort towards establishment of the NIEO.[42] The Declaration refers to the NIEO movement as being 'one of the principal guarantees for the creation of better conditions for all peoples to reach a life worthy of human dignity'.[43] It enlists the cooperation of all members of the international community 'on the basis of sovereign equality and removal of the disequilibrium that exists between them'.[44] Some familiar principles of international law have been restated in the language of rights.[45] Amongst the main principles[46] of the NIEO enunciated in the Declaration, the only direct reference to human rights is to the right of self-determination of all peoples[47] and the right of 'peoples of territories which are under colonial and alien domination and foreign occupation to achieve their liberation and to regain effective control over their natural resources and economic activities'.[48]

With respect to economic development, the Declaration diagnoses that the prevailing disparities between countries in the world is because of the economic problems that plague the developing countries[49] and that accelerated development through comprehensive international economic cooperation[50] and structural economic changes would help overcome the inequalities and remove the disequilibrium between States.[51] The structural economic changes it recommends are a just and equitable relationship between prices of raw materials and finished goods, the improvement of the competitiveness of raw materials with respect to synthetic goods, preferential treatment for developing countries in trade, access to technology, support for producers' associations and the provision of development assistance free of political and military conditions.[52]

One of the principles listed in the Declaration refers to the notion of development. It urges developing countries 'to concentrate all their resources for the cause of development'.[53] There is no elaboration on the notion of development in the Declaration. It supports not only the economic wellbeing of all peoples, including the present and the future generations, but also their social wellbeing. Its aim is to address disparities in economic growth so as to ensure social development, peace and justice[54] for all, and not just economic wellbeing. When this is read along with the final section of the Declaration, which aspires to the creation of better conditions for all peoples to live in dignity, a particular idea of development can be derived from the Declaration. Development is the notion of social, economic and political wellbeing of people, through the development of their economies, to be achieved with the aid of the NIEO. The reference to human dignity qualifies the idea of development and introduces the notion of a 'minimum' required to satisfy the achievement of human dignity. Such an understanding of development is also consistent with the expressions on development in the resolution on the second Development Decade, which refers to provisions for minimum standards of living through ensuring health care, water supply security, food security, population control and management, employment opportunities and the preservation of the environment.[55]

In expressing the view that international assistance and preferential treatment of States are in keeping with the premises of the modern welfare state, Oscar Schachter supported the conception of the NIEO movement as being instrumental to satisfy the basic human needs of disadvantaged communities.[56] However, no obligation exists in the international community, at the present time, similar to national obligations relating to welfare or social security provisions.[57] Of the host of issues that Jack Behrman raises as important for the NIEO, most relate to basic human needs (the plight of Palestinian refugees, the ghettos around the world, famine, the prices of necessities, loss of pride) and the means to support them (unjustifiable income disparities, the presumed exploitative power of large corporations, boredom in or dehumanisation of work),[58] supporting the view that the NIEO fundamentally exists to serve human needs. Evaluating the contribution of the Charter on Economic Rights and Duties of States, one commentator states that the rationale of the NIEO is development – 'the final goal of development, on the other hand, is an ever increasing welfare of the individual, provided the benefits are shared by all'.[59]

The Declaration not only incorporates a minimalist idea of development, in order to ensure that human dignity is protected, it also aims at facilitating conditions required for the further development of peoples. The reference to the gap between developing and developed countries could still exist even after the minimal development needs of peoples are fulfilled. The Declaration seeks to bridge this gap through the establishment of an equitable economic order.[60] The 'solid potential for improving the well-being of all peoples'[61] through the distribution of benefits that accrue from such progress can be utilised to fulfil human dignity and also to enhance the lifestyle and luxury of all peoples. The developed countries lodged many fundamental reservations about the NIEO for the reason that the idea of development was more than just minimalist; it could be interpreted as an open-ended concept.

Full and effective participation on the basis of equality of all countries is an important part of the movement for economic reform pushed forward by the developing countries. Reference to the Cold War and the use of economics as an instrument to achieve its ends are included in the Declaration which asserts the 'right of every country to adopt the economic and social system that it deems the most appropriate for its own development and not to be subjected to discrimination of any kind as a result.'[62] The effort to distance themselves from colonial policies on raw materials played a major role in the formulation of the principles of the Declaration, and the States established the right to exercise permanent sovereignty over their natural resources. The Declaration also recognised the need to regulate and supervise 'the activities of transnational corporations by taking measures in the interest of the national economies of the countries where such transnational corporations operate'.[63]

4.2.5 The Charter of Economic Rights and Duties of States (1974)

The purpose of the Charter of Economic Rights and Duties of States is 'the establishment of the new international economic order, based on equality, sovereign equality, interdependence, common interest and cooperation among all states, irrespective of their economic and social system'.[64] The provisions of the Charter seek to achieve substantial equality between sovereign States – not only formal equality granted under international law. It uses the language of rights and duties and seeks to strengthen the cooperation between States in order to bring about substantial equality. Such language in the Charter is an unusual proclamation by the States,[65] but one that points to the human rights connotations of economic development. The reference to rights in the Charter can be viewed as an aggregate of the rights of humanity; the use of the term 'duties' can be view in a similar way.

Much of the Charter reflects the ideas and principles embodied in the Declaration on the Establishment of the New International Economic Order and the Programme of Action because it was adopted soon after the two landmark resolutions.[66] The Charter also contains provisions for human development. Article 7 holds States responsible for the 'economic, social and cultural development of its people'. Under Article 14, all States share the responsibility of improving the 'welfare and living standards of all peoples, in particular, those of developing countries'. This article provides an explicit link between peoples' development and economic development. It relates the progressive dismantling of obstacles to trade to improving living standards of all peoples, particularly those living in developing countries. It also restates some of the other NIEO objectives such as equitable and remunerative pricing, diversification of exports, and accelerated rate of growth in order to bring about better standards of living of people.

The General Assembly resolutions on the NIEO, including the Charter initiated by the Third World was a result of deteriorating economic conditions leading to poverty.[67] Perhaps 'developing countries chose to focus on particularly ill-designed measures to translate their objectives into reality',[68] and consequently there has been an increase in the number of least-developed countries since the establishment of UNCTAD.

4.2.6 The Convention on the Law of the Sea's CHM regime as NIEO in action

The developing States saw an opportunity in the formulation of the seabed mining rules – the creation of unique operating principles based on their vision for a new economic order for the purpose of governing deep seabed natural resources. The 1994 Convention on the Law of the Sea is a unique regime, allowing for fair and equitable distribution of resources from the seabed, even if it is much less radical than its original formulation. The then Secretary-General of the UN, Kurt Waldheim, speaking at the opening session of the Third Conference on the Convention on the Law of the Sea, clearly posited the link between the movement for equitable economic systems and the equitable management of resources from the seabed. He stated:

> Finally, and most important in bringing about the Conference, had been the mounting pressure on world resources and the awareness that the sea-bed and the oceans contained some of the largest unexploited reserves available to man. The Conference had been convened because of the realisation that those resources must be developed in an orderly manner for the benefit of all and must contribute to a more equitable and workable global economic system.[69]

The obligation of States to support and aid the development of populations outside of their national jurisdictions is one of the NIEO principles. This becomes necessary because of the inequalities prevailing in the global economic system. The CHM concept, by focusing on the notion of mankind or humanity, embodies this principle of obligations beyond the borders. The system of site-banking, providing training and equitable market prices for technology, anti-monopoly clauses and other arrangements in the Convention are designed to benefit people who might otherwise be left out of the profits accruing from seabed mining.

States referred to the development of people as the main goal of the CHM regime in the General Assembly debates on governance of the oceans. Introducing the concept as a means of ocean governance, the representative of Malta, Ambassador Pardo stated:

> The use of the seabed and of the ocean floor, underlying the seas beyond the limits of present national jurisdiction, and their economic exploitation shall be undertaken with the aim of safeguarding the interests of mankind. The net financial benefits derived from the use and exploitation of the seabed and of the ocean floor shall be used primarily to promote the development of poor countries.[70]

Belgium referred to a resolution adopted by the Economic and Social Council[71] and stated that the CHM regime would raise the economic level of peoples throughout the world, especially those living in the developing countries.[72] El Salvador spoke for the establishment of a regime that would allow 'equitable participation of all peoples in the utilisation of the seas'.[73] While most references were made in relation to countries, about equitable sharing between developed and developing, between seafaring and landlocked countries, occasional references were made, casting aside the veil of statehood, directly to the peoples. The representative of Sri Lanka expressed the opinion that a CHM regime led to 'all peoples receiving their legitimate share of the benefits, and that the resources were exploited rationally and without damage to the environment'.[74] As the discussions progressed and the jargon became more economic, it still retained the focus on the individual on whose rights and abilities the economic system would have an effect. The human-centred development notion, which was introduced by Pardo in the General Assembly in 1967, continuously underpinned the proceedings at the Conference.

4.3 The RTD and the NIEO

The RTD is a human right that is relevant to economic development and human development, both of which are interrelated. It was formulated by the countries in the South and was formed during the period of the decline of the NIEO movement. It was believed that the attainment of political independence (through the assertion of the right to self-determination), with provisions for the rule of law and human rights, would bring about freedom from deprivation, extreme poverty and hope that a better standard of living could be attained, but that proved not to be the case.[75] Hence the RTD is an attempt to articulate more clearly and precisely the violations of human rights caused by poverty and deprivation. It has gained momentum since then and is applicable in the context of the human genome and governance issues. In this section, the link between NIEO and the RTD will be examined. The epistemology and political development of the right will be analysed and the steps taken for its implementation will also be investigated.

The right to health, subject to certain qualifications, is a key part of the implementation of the RTD. It will be argued that governance of the human genome should have as its central aim the fulfilment of the right to health and thereby contribute to the RTD. A CHM regime for the human genome would bring health benefits for all peoples, and necessitate the equitable allocation of health research resources and health care resources in order to benefit the peoples of the world. A CHM regime will have implications for the way research ethics committees around the world function, and the stakeholders they take into consideration when making decisions on applications.[76] Global justice and health equity will play a significant role in their decision-making process.

When the human rights movement gained momentum towards the end of the Cold War, the push towards development was cast in human rights terms and embodied in the RTD. It was however opposed by many developed countries[77] who saw it as a means to resuscitate the NIEO.[78] However, the demands of the RTD are much more modest. It will be shown in the next chapter that it is also a moral right. The 1986 Declaration on the Right to Development established the RTD as a human right and it alluded to the promotion of efforts towards the establishment of a new international economic order in its Preamble. Article 3(3) of the Declaration states that 'States should [...] fulfil their duties in such a manner as to promote a new international economic order based on sovereign equality, interdependence, mutual interest and co-operation among all states, as well as to encourage the observance of human rights'. The RTD was reiterated by the 1993 Universal Declaration of Human Rights as a right to a certain state of being, which enables the individual to enjoy the benefits arising out of development.

By its emphasis on the health of individuals, the 1997 Declaration on the Human Genome and Human Rights can be interpreted as being instrumental in the implementation of the RTD. The 1986 Declaration on the Right to Development identifies solidarity and international cooperation as being essential for the fulfilment of the RTD and a similar emphasis is placed on cooperation, in order to address health concerns of 'large numbers of the world's populations'.[79] It is argued, in the final section, that the RTD occupies a place in the governance of the human genome since it is embodied in the 1997 and 2005 Declarations, through its link to the right to health.

4.4 The emergence of the RTD

The RTD is derived from the Charter of the UN and the various international human rights instruments. Through the Charter and the International Covenant on Economic, Social and Cultural Rights (ICESCR), the members of the UN undertook to promote higher standards of living, full employment and conditions of economic and social progress and development for all.[80] Philip Alston traces the RTD back to as far as May 1944, to the Declaration of Philadelphia, of the International Labour Organization, which affirms that:

> all human beings, irrespective of race, creed or sex, have the right to pursue both their material well-being and their spiritual freedom in conditions of freedom and dignity, of economic and social opportunity.[81]

He also quotes a commentator who expressed the view that 'the root right from which all others stem is the right to life – the right to the fullest and finest development of the potentialities of the human personality, in the framework of the common good.'[82] In the International Conference on Human Rights in Tehran in 1968, States agreed that 'The achievement of lasting progress in the implementation of human rights is dependent upon sound and effective national and international policies of economic and social development'.[83] It noted that the widening gap between the economically developed and developing countries impeded the realisation of human rights in the international community.[84] The RTD was first articulated by Kéba M'Baye in 1972.[85] Karel Vasak was also developing his theory of a third generation of rights around the same time.[86]

The General Assembly recognised the RTD in a 1979 resolution as a right that would bring about equality of opportunity. It stated that 'the right to development is a human right and [that] equality of opportunity for development is as much a prerogative of nations as of individuals within nations'.[87]

In subsequent resolutions, the General Assembly recommended that the UN Commission on Human Rights (UNCHR) 'promote the right to development as a human right'[88] and reiterated the RTD as 'an inalienable human right'.[89] The UNCHR took up the task of promoting the RTD and recommended a study on 'the international dimensions of the right to development as a human right in relation with other human rights based on international co-operation, including the right to peace, taking into account the requirements of the New International Economic Order and fundamental human needs'.[90] In keeping with the objective of achieving substantive equality of all peoples, the CHR recommended further study on 'the regional and national dimensions of the right to development as a human right, paying particular attention to the obstacles encountered by developing countries in their efforts to secure the enjoyment of that right'.[91] In 1981, a Working Group of 15 governmental experts was established by the CHR to prepare a draft instrument on the RTD.[92]

4.4.1 The Declaration on the Right to Development

The General Assembly adopted the Declaration on the Right to Development in 1986.[93] It recognises the human person to be the central subject of the development process and that development policy should therefore make the individual the main participant and beneficiary of development. Article 1 of the Declaration defines the RTD as:

> an inalienable human right by virtue of which every person and all peoples are entitled to participate in, and contribute to, and enjoy economic, social, cultural and political development, in which all human rights and fundamental freedoms can be fully realised.[94]

This echoes the concept of development expounded during the NIEO period, which viewed development as 'the realisation of the potentialities of the human person in harmony with the community'.[95] It reflects the central theme of NIEO, which is international cooperation and assistance, and can be seen as an attempt to crystallise international cooperation into legal obligations.[96] The primacy and indivisibility of civil, political, economic, social and cultural rights are expressed in the formulation of the RTD. The Declaration seeks to integrate the so-called 'negative' and 'positive' rights and attempts to reframe the division between human rights. Although human rights are not easily prone to a hierarchical ordering, the concept of the RTD seeks a fundamental status for itself, via the qualification 'inalienable'.

The RTD is a human right which is essential for the enjoyment of all other human rights and fundamental freedoms. It refers to the full development of the individual and therefore can be 'very properly a fundamental human right – a basic one, which at the same time conditions and implies the right to development of developing states and peoples'.[97] The RTD of countries, although acceptable as a group (aggregate of individuals) right,[98] is still best described as a human right, which belongs to individuals. In this sense, the reference to States as right holders in the declaration is acceptable. As Georges Abi-Saab argues, there might be some value in viewing it as a collective right, 'as a matter of legal policy', if much more can be achieved for the individual 'by considering the right to development as a collective right.'[99] But a further derivation that the RTD belongs to developing States and peoples only is flawed. If classified as a human right, it is a right vested in everybody – whether they exercise that right is a different issue. The developed States and peoples possess the RTD in the same way as the developing States and their peoples and could claim it if at any point in time they were in a situation where they required to.

The right itself is said to include:[100]

- full sovereignty over natural resources[101]

- self-determination[102]
- popular participation in development[103]
- equality of opportunity[104]
- the creation of favourable conditions for the enjoyment of other civil, political, economic, social and cultural rights.[105]

Although these elements can be gleaned from the Declaration, there were concerns that a fixed set of criteria would entrench a notion of development, which in itself is a problematic term.[106] This was a concern about the NIEO on the part of some commentators which continues to persist with respect to RTD.

Article 1(2) states that the human RTD implies 'the full realization of the right of peoples to self-determination', and includes 'the exercise of their inalienable right to full sovereignty over all their natural wealth and resources'. While freedom from subjugation and the right to own property can be justified as human rights, the exercise of full sovereignty over natural resources and the right to self-determination should be seen as extensions of the human rights to property and freedom. The encouragement for popular participation in the development process stems from the need for people from developing countries and former colonies to self-govern. It enables the exercise of autonomy by empowering individuals. It is an integral part of the RTD. Article 8 of the Declaration states:

> States should undertake, at the national level, all necessary measures for the realization of the right to development and shall ensure, *inter alia*, equality of opportunity for all in their access to basic resources, education, health services, food, housing, employment and the fair distribution of income.

The right against discrimination is a well-established human right, commonly classified as a negative right. In the above section on equality of opportunity, States and other individuals have a positive duty to bring about equality, not just in formal terms. The Declaration requires the 'creation of favourable conditions for the enjoyment of social and economic development'. States have the 'duty to formulate appropriate national development policies that aim at the constant improvement of the well-being of the entire population and of all individuals'.[107] They are urged to take:

> resolute steps to eliminate the massive and flagrant violations of the human rights of peoples and human beings affected by situations such as those resulting from apartheid, all forms of racism and racial discrimination, colonialism, foreign domination and occupation, aggression, foreign interference and threats against national sovereignty, national unity and territorial integrity, threats of war and refusal to recognize the fundamental right of peoples to self-determination.[108]

The NIEO could be taken as the blueprint for the evolution of the RTD;[109] it is certainly reflected in the above provisions. Special mention is made of women, who could be left out of the development process, and States are urged to enact appropriate reforms to eradicate social injustices:[110] 'States have the duty to co-operate with each other in ensuring development and eliminating obstacles to development'.[111] Article 4 reinforces this duty by stipulating that States should 'take steps, individually and collectively, to formulate international development policies with a view to facilitating the full realization of the right to development'.

Inequality amongst peoples is recognised as the status quo in the world, and the Declaration advocates 'sustained action to promote more rapid development of developing countries'.[112] It sanctions the special status for developing countries based on need.[113] In keeping with its quality as an inalienable right, the Declaration concludes by stating that steps should be taken for the 'full exercise and progressive enhancement of the right to development'.[114] Although similarly worded to the ICESCR, it differs from it because the RTD does not envisage progressive realisation but progressive enhancement.

However, being enshrined in a resolution of the General Assembly, the Declaration does not create binding obligations, similar to those arising from the ICESCR. It could be argued though, that the Declaration does not create any new rights, but reinforces the two human rights covenants. It creates layers within a particular right and bestows an inalienable quality to some of the layers, hence stratifying the rights in order that the individual be able to attain a minimum standard of living.[115]

The indivisibility and interdependence of human rights are restated and the unequal treatment of ESCR is subtly pointed out through the provision that 'equal attention and urgent consideration should be given to the implementation, promotion and protection of civil, political, economic, social and cultural rights' on the part of States.[116] The maintenance of international peace and security, a prerequisite for fulfilment of human rights, underpins the objectives of the Declaration and to that end all States are urged to 'do their utmost to achieve general and complete disarmament under effective international control'.[117] Redistribution of wealth, if it is required for the achievement of the RTD of all peoples, is implicit in the Declaration. Supplementing this objective, the Declaration elaborates on resource allocation in Article 7. It connects military expenditure to development and prioritises the latter over the former, however it must be mentioned that only one component of military expenditure is cited. The Declaration states that 'resources released by effective disarmament measures are to be used for comprehensive development, in particular that of the developing countries'. The issues of reallocation and distribution have been at the heart of the development debate.[118]

4.4.2 The Vienna Declaration and Programme of Action (VDPA) and the RTD

The 1993 VDPA, adopted at the World Conference on Human Rights, provided a 'unique opportunity' for a 'comprehensive analysis of the international human rights system and of the machinery for the protection of human rights'.[119] It reaffirmed the RTD as a universal and inalienable right and an integral part of all fundamental human rights.[120] The World Conference was a culmination of a series of regional conferences aimed at achieving a reappraisal of human rights law and policy. It reiterated the provisions of the 1986 Declaration in Articles 10 to 14.

A landmark statement on the unequal status of various human rights was adopted. On the understanding that all human rights are universal, indivisible, interdependent and interrelated, States renewed their commitment to 'treat human rights globally in a fair and equal manner, on the same footing, and with the same emphasis'.[121] The following concerns formed the main part of the VDPA: poverty, health, solidarity, increased international cooperation, removal of economic obstacles that hinder development of poor countries, burden of debt alleviation and special measures for vulnerable groups (women, children, migrant workers, refugees, minorities, indigenous people) to enable them to fulfil their rights to health, food, and freedom of poverty. Apartheid, genocide, torture, racial discrimination, ethnic cleansing, enforced disappearances, systematic rape of women in conflict situations, and other gross violations of human rights were also addressed.

The stated objective of the VDPA is:

> to fulfil the aspirations of all the peoples for an international order based on principles enshrined in the Charter of the United Nations, including promoting and encouraging respect for human rights and fundamental freedoms for all and respect for the principle of equal rights and self-determination of peoples, peace, democracy, justice, equality, rule of law, pluralism, development, better standards of living and solidarity.[122]

Article 9 of the VDPA prioritises the least-developed countries in Africa as needing the support of the 'international community in order to succeed in their transition to democracy and economic development'. Read in conjunction with the provisions on solidarity to 'achieve substantial progress in human rights by an increased and sustained effort of international cooperation and solidarity'[123] and to facilitate equitable 'economic relations and a favourable economic environment at the international level',[124] the VDPA supports positive obligations on the part of all countries to enable the fulfilment of the RTD.

The Declaration prioritises the eradication of extreme poverty. Article 14 states 'the existence of widespread poverty inhibits the full and effective enjoyment of human rights; its immediate alleviation and eventual elimination must remain a high priority for the international community'. This section closely resembles the language in the 1986 Declaration, which urges States to fulfil the RTD so that an individual can participate in and contribute to all aspects of development. While the VDPA requires that poverty be eradicated so that an individual can enjoy fundamental freedoms and human rights, such a condition is also essential to participate in, and contribute to, development, while the fulfilment of other freedoms such as liberty, autonomy and participation are also necessitated for the enjoyment of the RTD. Dignity and poverty do not go together. Poverty creates extremely difficult conditions for dignity to thrive in and leads to violations of human dignity. Article 25 continues on the grave issue of poverty eradication and protection of human dignity, stating that:

> The World Conference on Human Rights affirms that extreme poverty and social exclusion constitute a violation of human dignity and that urgent steps are necessary to achieve better knowledge of extreme poverty and its causes, including those related to the problem of development, in order to promote the human rights of the poorest, and to put an end to extreme poverty and social exclusion and to promote the enjoyment of the fruits of social progress. It is essential for States to foster participation by the poorest people in the decision-making process by the community in which they live, the promotion of human rights and efforts to combat extreme poverty.

States are urged to take action in cases of children in extreme poverty and the declaration also places particular priority 'on reducing infant and maternal mortality rates, reducing malnutrition and illiteracy rates and providing access to safe drinking water and to basic education'.[125]

The VDPA professes the belief that 'lasting progress towards the implementation of the right to development requires effective development policies at the national level, as well as equitable economic relations and a favourable economic environment at the international level'.[126] The right to life and health for everyone are important human rights and the VDPA notes the right of everyone to enjoy the benefits of scientific progress and its applications. It makes particular mention of advances in biomedical sciences and states that it 'may have potentially adverse consequences for the integrity, dignity and human rights of the individual, and calls for international cooperation to ensure that human rights and dignity are fully respected in this area of universal concern'.[127] The breadth and scope of the VDPA reinforces the value and relevance of ESCR within the scheme of human rights and the role of human rights in upholding fundamental values in society.[128]

NIEO, the Declaration on the Right to Development and the UN Charter have led to the evolution of the understanding of development as a human right. From an open-ended interpretation of development, the meaning and content of the RTD have become narrower as further formulations have been made. The VDPA focuses and sharpens the right, with its references to extreme poverty and deprivation giving a context to the interpretation of the RTD.

4.4.3 The High Commissioner for Human Rights and the RTD

The international community continued to highlight the importance to the RTD through the establishment of a new position of High Commissioner for Human Rights with a mandate to promote the RTD.[129] The General Assembly passed a resolution in December 1993 which, while affirming its commitment to the VDPA, recommended that the UN Secretary-General to appoint a High Commissioner who was to 'recognize the importance of promoting a balanced and sustainable development for all people and of ensuring realization of the right to development, as established in the Declaration on the Right to Development',[130,131] The General Assembly made two important statements through Resolution 48/141. First, it made the promotion and the protection of all human rights one of the priorities of the international community.[132] Second, it stated that the RTD is the most important human right that should be fulfilled and protected as a matter of priority by the international community.

4.4.4 Implementation of the RTD

In 1998, the UNCHR adopted a resolution establishing the Open-ended Working Group on RTD and an Independent Expert on the RTD[133]. The RTD Working Group was tasked with monitoring and reviewing the promotion and implementation of the RTD by reviewing reports submitted by States and non-governmental organisations (NGOs), and providing a sessional report to the CHR. The Independent Expert was to undertake studies for the Working Group on the progress towards implementation. In 2002, the UNCHR passed a resolution on the RTD,[134] taking note of the UN Millennium Declaration adopted by the General Assembly on 8 September 2000. It stressed the need to broaden and strengthen the 'participation of developing countries and economies in transition in international economic and decision-making and norm-setting'. The resolution continued the trend of specifying projects and programmes with the potential to realise the RTD: the fight against HIV/AIDS, the New Partnership for Africa's Development and the UN Millennium Declaration. Goal 8 of the Millennium Declaration aimed at forging partnerships for development. A high-level task force was established to suggest criteria on the basis of which periodic evaluations could be made to determine the progress achieved for the fulfillment of this goal. The task force was significant because it brought together experts from finance and trade institutions.[135]

The discussions during the 7th Session of the Working Group (Geneva, 9–13 January 2006[136]) welcomed the work of the task force and the focus on development partnerships, but expressed concern that the focus was more at the national level rather than on developing international partnerships.

The Non-Aligned Movement, represented by Malaysia, noted with concern the absence of tangible progress in the operationalisation of the RTD in the two decades that had passed since the adoption of the 1986 Declaration. It suggested that the Working Group focus its deliberations on developing a legally binding instrument on the RTD.[137] Austria, on behalf of the European Union (EU), acceded and candidate countries Norway, the Republic of Moldova and Ukraine welcomed the constructive dialogue over the past two years which had shifted the debates on the Working Group from the general to the specific and from the conceptual to the operational.

The Working Group also underlined the importance of increasing financial assistance to developing countries; the need to mobilise resources to combat pandemics such as HIV/AIDS, malaria and tuberculosis; the central role of foreign direct investment in improving the effectiveness of global partnerships; the impediment of the debt burden to achieving the RTD; the need for additional resources to finance efforts towards the achievement of Millennium Declaration Goals (MDGs); and the need for States to honour their commitment to dedicate 0.7 per cent of their gross national product (GNP) to official development assistance (ODA).[138]

Several delegations underlined that the RTD was a unique right, and that it was this right, rather than merely a human rights-based approach to development, that the Working Group should seek to mainstream and make operational.[139] The report of the high-level task force was seen to focus excessively on national rather than international responsibilities of States.

The newly formed Human Rights Council adopted a resolution supporting a move towards the formulation of 'an international legal standard of a binding nature, guidelines on the implementation of the right to development and principles for development partnership'.[140]

4.5 The RTD and health

The tensions between the implementation of the RTD and the unwillingness on the part of some States to recognise such a right stem from the ambiguity involved in defining the concept. This has been addressed by the Independent Expert on the RTD by way of his reports and publications. In his seminal article in 2002, Dr Arjun Sengupta defines the RTD as:

> a process of development which leads to the realisation of each human right and all of them together and which has to be carried out in a manner known as rights-based, in accordance with the international human rights standards, as a participatory, non-discriminatory, accountable and transparent process with equity in decision-making and sharing of the fruits of the process.[141]

The RTD has been interpreted as both a right to a set of 'basic' rights and a right to a process.[142] It seems to contain 'an objective and the means to it'.[143] In most conceptions of the RTD, the indivisibility of human rights is stressed. This has drawn criticism from various authors who are of the opinion that the RTD is nothing but a 'synthesis of more traditional human rights'.[144]

In trying to isolate elements of the RTD, the Independent Expert has listed the right to food, health and primary education as the three basic rights that have to be 'immediately addressed' in order for the RTD to be fulfilled. The emphasis on health is not novel. In 1978, the Declaration of Alma-Ata proclaimed that 'The existing gross inequality in the health status of the people particularly between developed and developing countries as well as within countries is politically, socially and economically unacceptable'. It went on to state that 'the promotion and protection of the health of all the people is essential to sustained economic and social development and contributes to a better quality of life and world peace'.[145]

The World Health Organization passed a resolution in 1979 called Health and the New International Economic Order,[146] urging both developed and developing States to put their efforts into the transfer of appropriate technology, resources and technical cooperation for the implementation of the primary health care approach towards improved health for all.[147] In a resolution passed by the General Assembly in November 1979 entitled Health as an Integral Part of Development,[148] States endorsed the Alma-Ata Declaration and declared that 'a combination of promotive, preventive, curative and rehabilitative measures constituted the key to the ultimate achievement of a healthful society', noting that 'a substantial portion of the population in many countries, developing as well as developed, lacks access to basic health services, and that people lacking adequate health cannot fully participate in or contribute to the economic and social development of their nation'. The commitment to health was renewed by the international community through the MDGs in 2000.[149] Although not worded in human rights terms, there is some support for human rights-based approaches to these goals.[150]

The Independent Expert recommended that the positive obligations of the international community to extend cooperation be discharged through the formation of 'development compacts', in which developing countries would undertake to fulfil their national human rights obligations, while the international community would provide resources and share costs. These steps are to be taken in addition to the other measures taken to fulfil and protect human rights under other international instruments. Such an arrangement has positive resonances with scholars who see it as an expression of 'our citizen-ness and humanity'.[151] The underlying structural problems are also the subject of the recommendations. A range of economic and trade measures that need to be carried out in the areas of trade; access to markets; debt adjustment for the poorest countries; transfer of resources and technology, protection of migrants and labour standards; and the restructuring of the international financial system to give the developing countries a greater share in power and decision-making and to increase the flow of private capital to their economies – all seem very similar to the manifesto of the NIEO.

Parallels can be drawn between the 1986 Declaration on the Right to Development and the 1997 Universal Declaration on the Human Genome and Human Rights. The 1986 Declaration was drafted at a time when developing countries perceived that their economies and peoples were sidelined in the development process; the 1997 Universal Declaration and the provisions relating to the CHM are also underpinned by concerns about the exacerbation of inequalities based on health and technology.

International cooperation and principles based on solidarity is a key feature of both documents. Equitable benefit sharing is a theme in the 1997 Universal Declaration and fair distribution of benefits is one of the objectives of the 1986 Declaration. The 1986 instrument urges States to complement the efforts of developing countries and participate in effective international cooperation in providing developing countries with appropriate means and facilities to foster their comprehensive development.[152] The 1997 instrument urges States to foster solidarity between population groups and research 'on the identification, prevention and treatment of genetically-based and genetically-influenced diseases, in particular rare as well as endemic diseases which affect large numbers of the world's population.'[153] Articles 17 to 19 of the 1997 Declaration seek to deepen and strengthen cooperation in the areas of health research, health care, technology and knowledge transfer, and the free exchange of scientific knowledge and information in medicine.

The 2005 Universal Declaration on Bioethics and Human Rights, when read in conjunction with the 1997 Universal Declaration, strengthens the solidarity and cooperation provisions. It considers it desirable to develop new approaches to social responsibility to ensure that progress in science and technology contributes to justice, equity and to the interests of humanity.[154] Articles 14 and 15 on social responsibility and health and the sharing of benefits attach importance to international cooperation in health and the objectives in health closely correlate with the MDGs.[155]

While the 1986 Declaration refers marginally to the reallocation of resources in the context of development,[156] the 1997 Universal Declaration can be interpreted to include issues of reallocation of resources within its section on 'Conditions for the Exercise of Scientific Activity'. Article 13 points to the 'responsibilities inherent in the activities of researchers' in the 'utilization of their findings' and their 'ethical and social implications'. The 1997 Universal Declaration identifies public and private science policy-makers as having particular responsibilities in this respect. Its following provision urges States to take appropriate measures in assessing the economic implications of genome research. A critical analysis of the role of intellectual property rights and its links with human rights would also have implications for resource allocation.

Notes

1 Declaration on the Right to Development, UN Doc. A/RES/41/128, 1986.

2 See 'The Bay of Pigs Invasion' in L. Freedman, *Kennedy's Wars: Berlin, Cuba, Laos, and Vietnam*, New York: Oxford University Press, 2000 and 'The Cuban Missile Crisis' in M. Frankel, *High Noon in the Cold War: Kennedy, Krushchev, and the Cuban Missile Crisis*, USA: Presidio Press, repr. 2005.

3 See talks given at the seminar on the 'Effects of Existing Unjust International Economic Order on the Economies of the Developing Countries and the Obstacle that This Represents for the Implementation of Human Rights and Fundamental Freedoms', Geneva, Switzerland, 30 June–11 July 1980, UN Doc. ST/HR/SER.A/8, 1980.

4 R. Looney, *New International Economic Order*, in R. J. B. Jones (ed.), *Encyclopedia of International Political Economy*, London: Routledge, 1999.

5 S. Chauhan, *GATT to WTO: Gandhian Alternative to New International Economic Order*, New Delhi: Deep and Deep Publications Pvt. Ltd, 2001, p. xvii.

6 Jack Behrman argues that 'the rules employed in the post war period 1945–70 applied only within a highly constrained system, one known as "neo-classical" and one which has not been practiced *fully* at any time and seldom outside of the Western World'. See J. N. Behrman, *Toward a New International Economic Order*, Paris: The Atlantic Institute for International Affairs, 1974, p. 9.

7 'Human Rights Approaches to Development', *NGLS Roundup 30*, November 1998, 1–6 at 4; P. Henriquez, *Perspectives on the New International Economic Order*, in R-J. Dupuy, *The Right to Development at the International Level*, Workshop, The Hague, 16–18 October 1979, The Hague Academy of International Law, Alphen aan den Rijn, The Netherlands: Sijthoff and Noordhoff, 1980, p. 185.

8 J. Stanovnik, *Towards the New International Economic Order*, Belgrade: Jugoslovenska Stavarnost, 1979, p. 12.

9 M. Bedjaoui, *Towards a New International Economic Order*, Paris: UNESCO, 1979, p. 20.

10 H. G. Johnson, *The New International Economic Order*, Graduate School of Business (Selected Papers No. 49), University of Chicago, 1976, p. 18

11 Ibid.

12 United Nations Development Decade: A Programme for International Economic Co-operation (I), UN Doc. A/RES/16/1710, 1961; United Nations Development Decade: A Programme for International Economic Co-operation (II), UN Doc. A/RES/16/1715, 1961.
13 Preamble, UN Doc. A/RES/16/1710, 1961.
14 Ibid., Section 1.
15 Ibid., Section 4(d): 'In a report issued in 1969, Secretary-General U Thant noted that the slower progress in development had been accompanied by the emergence or aggravation of major imbalances that imperilled future growth. Without greater progress in food production and the more effective control of communicable diseases, the necessary conditions for steady economic and social development could hardly be said to have been created. At the same time, the Secretary-General pointed out, the experience of a few countries had demonstrated that "given a favourable constellation of circumstances and policies, an adequate and sustained pace of development can be achieved," and acceptance of development as a fundamental objective had gradually wrought a desirable change in attitudes and modes of action on the part of developing countries.' *First UN Development Decade*, available online at http://www.nationsencyclopedia.com/United-Nations/Economic-and-Social-Development-FIRST-UN-DEVELOPMENT-DECADE.html (accessed 29 August 2006).
16 Proclamation of Tehran, Section 12, Office for the High Commissioner of Human Rights, available online at http://www.unhchr.ch/html/menu3/b/b_tehern.htm (accessed 29 August 2006).
17 International Development Strategy for the Second United Nations Development Decade, UN Doc. A/RES/25/2626, 1970; Role of the Regional Economic Commissions in the Second United Nations Development Decade, UN Doc. A/RES/25/2687, 1970.
18 R. B. Lillich (ed.), *Economic Coercion and New International Economic Order*, Charlottesville, Virginia: MC Law Publishers, 1976, p. 78.
19 Declaration on the Establishment of a New International Economic Order, UN Doc. A/RES/S-6/3201, 1974.
20 Programme of Action on the Establishment of a New International Economic Order, UN Doc. A/RES/S-6/3202, 1974.
21 UN Doc. A/RES/29/3281, 1974.
22 UN Doc. A/RES/S-7/3362, 1975.
23 W. Brandt, *North South: A Programme for Survival*, USA: The MIT Press, 1980.
24 G. S. Varges, *The New International Economic Order Legal Debate: Background, Status, and Alternatives*, European University Studies Series II, vol. 326, Frankfurt am Main: Peter Lang, 1983, p. 2.
25 Ibid., p. 5.
26 T. Franck and M. Munansangu, *The New International Economic Order: International Law in the Making?*, New York: UN Institute for Training and Research, 1982.
27 Article 17, Charter of the UN, 1945.
28 *Texaco Overseas Petroleum Company/California Asiatic Oil Company and the Government of the Libyan Arab Republic (Arbitration)*, reproduced in *International Legal Materials* 17, 1, 1978, 1–37.
29 G. Corea, *Need for Change: Towards the New International Economic Order*, Oxford: Pergamon Press, 1980, pp. 61–142.
30 'ACP-EEC Lomé Convention', *The Courier* 31, Special Issue, March 1975.
31 Yaounde Convention, Arusha and Lagos Agreements, 1966 (not in force because France and Luxemburg did not ratify).
32 T. A. Aguda, *Human Rights and the Right to Development in Africa*, The Nigerian Institute of International Affairs, Lecture Series No. 55, 1989, pp. 4–5.
33 R. Jungnickel, G. Koopmann, *The Role of the Multinationals in the NIEO*, in *New International Economic Order: A Third World Perspective*, P. K. Ghosh (ed.), *Third World Development: a basic needs approach*, Connecticut: Greenwood Press, 1984, pp. 198–9.

34 The Convention on Biological Diversity confers sovereign rights to States on biological diversity found within the territory of the State. The Convention on Biological Diversity, reproduced in *International Legal Materials* 31, 1992, 822–32.

35 See A. G. Frank, *Rhetoric and Reality in the New International Economic Order,* School of Development Studies, University of East Anglia, Discussion Paper No. 35, 1983, p. 45; Ambassador E. O. Sanu, *The Lomé Convention and the New International Economic Order*, The Nigerian Institute of International Affairs, Lecture Series No. 18, 1979, p. 36.

36 D. French, '"From Seoul with Love" – the Continuing Relevance of the 1986 Seoul ILA Declaration on Progressive Development of Principles of Public International Law relating to a New International Economic Order', *Netherlands International Law Review* 55, 2008, 3–31.

37 Annotated Agenda of the Tenth Session of the United Nations Conference on Trade and Development, available online at http://www.unctad-10.org/index_en.htm 2000 (accessed 21 October 2003).

38 Ibid., pp. 22–5.

39 S. Amin, *After the New International Economic Order: The Future of International Economic Relations*, UNCTAD Trade and Development Report, 1982, pp. 9–23.

40 UN Doc. A/RES/S-6/3201, 1974, paragraph 2.

41 Ibid., paragraph 1

42 Ibid., paragraph 3.

43 Ibid., paragraph 6.

44 Ibid., paragraph 3.

45 M. G. Erasmus, *The New International Economic Order and International Organisations: Towards a Special Status for Developing Countries?*, Frankfurt/Main: Haag und Herchen, 1979, p. 88.

46 Section 4, UN Doc. A/RES/S-6/3201, 1974. The section is best classified as a 20-point manifesto which contains some principles.

47 Ibid., Section 4(a).

48 Ibid., Sections 4(h) and 4(i).

49 Ibid., Sections 4(b)–(c).

50 Ibid., Section 6.

51 Ibid., Section 3.

52 Economists, while recognising that the 'Third World countries unquestionably need resources and skills to move up and join the First World', believed that 'these resources should be transferred within the apparatus of a market system – through grants, aid and foreign investment – rather than through the distortion of commodity prices'. A. F. Ott, *The Impact of the New International Economic Order on the Third World* in W. G. Tyler, *Issues and Prospects for the New International Economic Order*, Toronto: Lexington Books, 1977, p. 23.

53 Section 4(r), UN Doc. A/RES/S-6/3201, 1974.

54 Ibid., Preamble.

55 Section 8 on Human Development, UN Doc. A/RES/25/2626, 1970.

56 O. Schachter, 'The Evolving International Law of Development', *Columbia Journal of Transnational Law* 15, 1976, 1–16.

57 C. W. Jenks, 'Economic and Social Change and the Law of Nations', *Receuil des Cours* 138, 1973, 455–502.

58 J. N. Behrman, *Toward a New International Economic Order*, Paris: The Atlantic Institute for International Affairs, 1974, pp. 68–9.

59 M. Sukijasovic, *The Contribution of the Charter on Economic Rights and Duties of States to the Elaboration of the Notion of "New International Economic Order"*, in *Legal Aspects of a New International Economic Order*, Belgrade: Institute of Comparative Law, 1979, pp. 44–5.

60 Section 1, UN Doc. A/RES/S-6/3201, 1974.

61 Ibid.
62 Ibid., Section 4(d).
63 Ibid., Section 4(g).
64 UN Doc. A/RES/29/3281, 1974. The US and 15 other States voted against the resolution. See L. Henkin, R. Pugh, O. Schachter and H. Smit, *International Law*, St. Paul, Minnesota: West Publishing Co., 2nd edn, 1987, p. 1054.
65 M. G. Erasmus, *The New International Economic Order and International Organisations: Towards a Special Status for Developing Countries?*, Frankfurt/Main: Haag und Herchen, 1979, p. 116. These rights and duties arise from the notion of sovereignty.
66 See Section VI, UN Doc. A/RES/S-6/3202, 1974, which states that the Charter of Economic Rights and Duties would be adopted at the 29th Regular Session of the General Assembly.
67 C. F. Bergsten in J. N. Bhagwati (ed.), *The New International Economic Order: The North-South Debate*, Cambridge, Massachusetts/London: The MIT Press, 1977, p. 347.
68 Ibid., p. 14.
69 *Third United Nations Conference on the Law of the Sea*, vol. I, New York: UN Publications, 1975, pp. 37–8.
70 UN Doc. A/RES/22/6695, 1967, at paragraph 3.
71 ECOSOC Doc. RES/40/1112, 1966.
72 UN Doc. A/RES/22/6695, 1967, at paragraph 45.
73 UN Doc. A/RES/27/8721, 1972, at paragraph 62.
74 *Third United Nations Conference on the Law of the Sea*, vol. I, New York: UN Publications, 1975, Summary Records of Meetings, First Committee, 2nd Meeting, paragraph 10.
75 T. A. Aguda, *Human Rights and the Right to Development in Africa*, The Nigerian Institute of International Affairs, Lecture Series No. 55, 1989, p. 13.
76 See Article 16, Universal Declaration on the Human Genome and Human Rights, UNESCO Doc. C/RES/29/16, 1997, which recommends the establishment of pluralist research ethics committees.
77 R. L. Barsh, 'The Right to Development as a Human Right: Results of the Global Consultation', *Human Rights Quarterly* 13, 1991, 322–38 at 322.
78 S. Marks, *Obstacles to the Right to Development*, François-Xavier Bagnoud Center for Health and Human Rights, 2003.
79 Article 17, Universal Declaration on the Human Genome and Human Rights, UNESCO Doc. C/RES/29/16, 1997.
80 Article 55, Charter of the UN, 26 June 1945: 'With a view to the creation of conditions of stability and well-being which are necessary for peaceful and friendly relations among nations based on respect for the principle of equal rights and self-determination of peoples, the United Nations shall promote: a. higher standards of living, full employment, and conditions of economic and social progress and development; b. solutions of international economic, social, health, and related problems; and international cultural and educational co-operation; and c. universal respect for, and observance of, human rights and fundamental freedoms for all without distinction as to race, sex, language, or religion.'
81 P. Alston, *The Right Development at the International Level*, in Rene-Jean Dupuy, *The Right to Development at the International Level*, Workshop, The Hague, 16–18 October 1979, The Hague Academy of International Law, Alphen aan den Rijn, The Netherlands: Sijthoff and Noordhoff, 1980, p. 100.
82 Ibid.
83 Article 13, *Proclamation of Tehran*, Office for the High Commissioner of Human Rights, available online at http://www.unhchr.ch/html/menu3/b/b_tehern.htm (accessed 29 August 2006).

84 Ibid., Article 12.

85 K. M'Baye, 'Le droit au développement comme un droit de l'homme', *Human Rights Journal* 5, 1972, 503–34; Keba Mbaye's Opening Remarks at the International Commission of Jurists Conference on Development, *Human Rights and the Rule of Law*, The Hague: Pergamon Press, 1981, p. 11.

86 T. A. Aguda, *Human Rights and the Right to Development in Africa*, The Nigerian Institute of International Affairs, Lecture Series No. 55, 1989, p. 14.

87 *Alternative Approaches and Ways and Means within the United Nations System for Improving the Effective Enjoyment of Human Rights and Fundamental Freedoms*, UN Doc. A/RES/34/46, 1979.

88 UN Doc. A/RES/35/174, 1980.

89 UN Doc. A/RES/36/133, 1981.

90 Commission on Human Rights RES/33/4, 1980.

91 Commission on Human Rights RES/37/4, 1979.

92 Commission on Human Rights RES/37/36, 1981.

93 UN Doc. A/RES/41/128, 1986.

94 Ibid., Article 1.

95 Secretary-General's Report E/CN.4/1334, 2 January 1979; see also C. Dias, *Development, Human Rights and the Rule of Law*, New York: Pergamon Press, 1981.

96 R. Zacklin, *The Right to Development at the International Level: Some Reflections on its Sources, Content and Formulation* in R. Dupuy, *The Right to Development at the International Level*, Workshop, The Hague, 16–18 October 1979, The Hague Academy of International Law, Alphen aan den Rijn, The Netherlands: Sijthoff and Noordhoff, 1980, p. 117.

97 H. G. Espiell, *Derecho International del Desarrollo*, University of Valladolid, 1975, p. 42.

98 R. Ferrero, Special Rapporteur of the Sub-Commission on Prevention of Discrimination and Protection of Minorities, *The New International Economic Order and the Promotion of Human Rights*, New York: UN, 1986, p. 27.

99 G. Abi-Saab, *The Legal Formulation of a Right to Development* in R. Dupuy, *The Right to Development at the International Level*, Workshop, The Hague, 16–18 October 1979, The Hague Academy of International Law, Alphen aan den Rijn, The Netherlands: Sijthoff and Noordhoff, 1980, pp. 163–4.

100 *The Content of the Right to Development*, Human Rights in Development, Office of the Commission on Human Rights, available online at http://www.unhchr.ch/development/right-02.html (accessed 29 August 2006).

101 Preamble and Article 1(2), Declaration on the Right to Development, UN Doc.A/RES/41/128, 1986.

102 Ibid., Article 1(2).

103 Ibid., Article 8(2).

104 Ibid., Article 8.

105 Ibid., Articles 3, 4, 5, 7 and 8.

106 S. Marks, A. Clapham, *International Human Rights Lexicon*, Oxford: Oxford University Press, 2005, p. 104.

107 Article 2(3), Declaration on the Right to Development, Article 1(2), UN Doc. A/RES/41/128, 1986.

108 Ibid., Article 5.

109 G. Abi-Saab, *The Legal Formulation of a Right to Development* in R. Dupuy, *The Right to Development at the International Level*, Workshop, The Hague, 16–18 October 1979, The Hague Academy of International Law, Alphen aan den Rijn, The Netherlands: Sijthoff and Noordhoff, 1980, p. 167.

110 Article 8, Declaration on the Right to Development, Article 1(2), UN Doc. A/RES/41/128, 1986.

111 Ibid., Article 3(3).

112 Ibid., Article 4(2).
113 S. R. Chowdhury, P. J. I. M. de Waart, *Significance of the Right to Development: An Introductory View*, in S. R. Chowdhury, E. M. J. Denters, P. J. I. M. de Waart, *The Right to Development in International Law*, Dordrecht: Martinus Nijhoff Publishers, 1992, p. 19.
114 Article 10, Declaration on the Right to Development, Article 1(2), UN Doc. A/RES/41/128, 1986.
115 See Chapter 5, 'Principle of Generic Consistency as the basis of the International Human Rights Order'.
116 Article 6, Declaration on the Right to Development, UN Doc. A/RES/41/128, 1986.
117 Ibid., Article 7.
118 M. Ul-Haq, *Reflections on Human Development*, Oxford: Oxford University Press, India, 2nd edn, 1999, p. 46.
119 Ibid., Preamble.
120 UN Doc. A/CONF.157/23, 1993.
121 Ibid., Article 5. See also ibid., Article 32: 'The World Conference on Human Rights reaffirms the importance of ensuring the universality, objectivity and non-selectivity of the consideration of human rights issues.'
122 Preamble, Vienna Declaration and Programme of Action, UN Doc. A/CONF.157/23, 1993.
123 Ibid.
124 Ibid., Article 10.
125 Ibid., Article 47.
126 Ibid., Article 10.
127 Ibid., Article 11.
128 P. Alston, *Making and Breaking Human Rights: The UN's Specialised Agencies and Implementation of the International Covenant on Economic, Social and Cultural Rights*, Human Rights and Development Working Paper Series, London: Anti-Slavery Society, 1979, pp. 42–4.
129 Part II, paragraph 18, Vienna Declaration and Programme of Action, UN Doc. A/CONF.157/23, 1993.
130 Ibid., paragraph 3(c).
131 The Human Rights Commission was replaced by the Human Rights Council in 2006 (UN Doc. A/RES/60/251, 2006). The new body assumes the role and responsibilities of the Commission on Human Rights relating to the work of the Office of the High Commissioner for Human Rights, as decided by the General Assembly in its Resolution 48/141. It could be argued that by making the human rights body an organ of the General Assembly its focus could move away from economic and social rights, and this would be detrimental to the realisation of the RTD. But it could also be argued that the General Assembly has been remarkably good at addressing social and development issues. The formulation of the Millennium Development Goals (MDGs) is one example of the international consensus that can be achieved in the General Assembly on issues that affect the realisation of economic and social rights.
132 The Resolution states that 'the promotion and the protection of all human rights is one of the priorities of the international community.'
133 Commission on Human Rights RES/72, 1998
134 Commission on Human Rights RES/69, 2002.
135 M. E. Salomon, 'Towards a Just Institutional Order: A Commentary on the First Session of the UN Task Force on the Right to Development', *Netherlands Quarterly of Human Rights* 23, 2005.
136 *Right to Development: Report of the Working Group on the Right to Development on its Seventh Session*, UN Doc. E/CN.4/2006/26, 2006.
137 Ibid., pp. 5–6.
138 Ibid., p. 6.
139 Ibid., p. 7.

140 UN Doc. A/HRC/Sub.1/58/L.30, 2006. Many countries including the US are opposed to legal commitments relating to the RTD. The US voted against the 1986 Declaration in the General Assembly and has consistently opposed any legally binding formulation of the right stating that it is committed to 'development'.
141 A. K. Sengupta, 'On the Theory and Practice of the Right to Development', *Human Rights Quarterly* 24, 2002, 837–89.
142 A. K. Sengupta, *The Right to Development as a Human Right*, François-Xavier Bagnoud Center for Health and Human Rights, Harvard School of Public Health, 1999.
143 A. K. Sengupta, 'On the Theory and Practice of the Right to Development', *Human Rights Quarterly* 24, 2002, 837–89 at 875.
144 Anne Orford quoting Jack Donnelly in *Globalisation and The Right to Development* in P. Alston (ed.) *Peoples' Rights*, Oxford: Oxford University Press, 2001, p. 140.
145 Declaration of Alma-Ata, International Conference on Primary Health Care, Alma-Ata, USSR, 6–12 September 1978.
146 WHO Doc. WPR/RC30.R13, 1979.
147 Ibid., Section 1.
148 UN Doc. A/RES/34/58, 1979.
149 UN Doc. A/RES/55/2, 2000.
150 P. Alston, 'Ships Passing in the Night: The Current State of the Human Rights and Development Debate Seen Through the Lens of the Millennium Development Goals', *Human Rights Quarterly* 27 (3), 2005, 755–829 at 806–7.
151 P. Uvin, *Human Rights and Development*, Connecticut: Kumarian Press Inc., 2004, p. 201.
152 Ibid., Article 4.
153 Article 17, Universal Declaration on the Human Genome and Human Rights, UNESCO Doc. C/RES/29/16, 1997.
154 Universal Declaration on Bioethics and Human Rights, UNESCO Doc. C/RES/33/36, 2005.
155 Also see Article 6, *Human Rights and Bioethics*, Commission on Human Rights Resolution RES/63, 1999: 'the importance of research on the human genome and its applications for the improvement of the health of individuals and mankind as a whole, to the need to safeguard the rights of the individual and his dignity, as well as his identity and unity, and to the need to protect the confidentiality of genetic data concerning a named person'.
156 Article 7, Declaration on the Right to Development; Article 1(2), UN Doc. A/RES/41/128, 1986.

5 Human rights, common heritage and development

A moral perspective

5.1 Introduction

Human rights law is a unique area of international law, in that its formation and development has been based on moral persuasiveness[1] much more than any other area of international law.[2] Human rights instruments mark a watershed in the development of public international law and have expanded its province.[3] They have revolutionised the notion of legal personality and transformed the way in which international organisations and institutions deal with persons who populate States.[4] The right to development (RTD) is a fundamental human right, which is as yet non-binding and non-justiciable. The content of the RTD is contested and most developed countries are not in favour of an open-ended notion of the RTD.

This chapter explores the moral aspects of the RTD through the Principle of Generic Consistency (PGC). The PGC is integral to a human rights theory elaborated by the philosopher Alan Gewirth. It is argued that the PGC, which is based on the generic needs of agency, provides a rational explanation to the existence of human rights in the modern world, and therefore that it should be considered as the necessary criterion of legal validity for the interpretation of human rights. Based on this premise, a particular definition for the RTD is built and it is demonstrated that moral obligations flow from it and, if adopted, legal obligations too.

5.2 Why Gewirth?

An exposition of the significant concepts in Gewirthian theory reveals the richness and versatility of this theory. The claims that Gewirth uses as the foundation of his theory are fundamental and extremely important at an empirical level. Action is at the centre of all life, at the molecular level and at the individual level. Erwin Schrödinger, the famous physicist, in his book 'What is Life?' noted that at the molecular level the cell must maintain itself in a state away from equilibrium since thermodynamic equilibrium is the very definition of death.[5] To stay away from a state of equilibrium, as Schrödinger states, it is essential that there be some movement and action. Action is essential for sustenance of all life forms. At the individual level, the right to life is also about the right to be able to act. The following passage illustrates the content of the right to life and the positive obligations associated with it:

> Under international human rights instruments, the assertion of the inherent right to life of every human being is accompanied by an assertion of the legal protection of that basic human right and of the *negative* obligation not to deprive arbitrarily of one's life (e.g., UN Covenant on Civil and Political Rights, Article 6 [1]; European Convention on Human Rights, Article 2; American Convention on Human Rights, Article 4 [1]; African Charter on Human and Peoples' Rights, Article 4). But this negative obligation is accompanied by the *positive* obligation to take all appropriate measures to protect and preserve human life. This has been acknowledged by the European Commission of Human Rights, whose case-law has evolved to the point of holding *(Association X versus United Kingdom* case, 1978) that Article 2 of the European Convention on Human Rights imposed on states also a wider and positive obligation "de prendre des mesures adéquates pour protéger la vie."[6]

The right to life is not just a negative right; it comes with the positive obligation to enable action or, in other words, the possibility of it. Action is the very basis of human existence and hence any theory that values action per se[7] is a theory that is worthy of attention.[8] The more fundamental the starting point, the more acceptable it could expect to become to diverse groups.

The notion of freedom, which is also fundamental in Gewirthian theory, is a value cherished in all societies. Even in instances where societies treat agents unequally, by placing restrictions on women in various ways[9], for example, the issue is not about the importance of freedom, but more about who should have it.

By appealing to logic, the principle of generic consistency captures a powerful human yearning to make sense of the world and its functioning through a certain type of ordering that is dictated by the brain through thought. Most societies also find logic appealing. Gewirthian theory taps into this fundamental urge to be logical. The PGC is based on a contingent argument that human rights exist and goes on to provide a hierarchy of rights, which has been in discussion for some time now within the field of human rights.[10] A theory which so explicitly advocates a hierarchy can be an important tool in urgently prioritising the fulfilment of human rights in the face of the increasing empirical evidence that large-scale violations of second-generation human rights are common occurrences.[11]

Human dignity is fundamental to international human rights law[12] and has gained renewed importance in the context of new genetics and biotechnology.[13] But its meaning is vague,[14] and it has been used as a 'place holder'[15] for other less presentable ideas. It is often seen as a notion that has to be clarified and explained in every context it is used, in order to steer away from an intuitive reference to the term. The Gewirthian theory of human rights provides us with a solid basis on which to conceptualise human dignity. As Gregory Walters puts it, 'the ultimate purpose of the rights (generic rights) is to secure for each person a certain fundamental moral status tied to agency, which is both the metaphysical and moral basis of human dignity'.[16]

5.3 Gewirthian theory

Alan Gewirth's endeavour is to show that morality is not a matter of personal choice or social convention but a fundamental necessity founded on rational analysis of human action and that it should be supported by the institutions we create to govern ourselves. Thus, 'human rights law is essentially non-utilitarian or counter majoritarian because it provides protection for individuals, groups and minorities so that in certain defined contexts their interests are not always sacrificed to those of the government or political majority of the day.'[17] Being based in human action, he appeals to all systems of governance which affect and shape human action. Although traditionally national institutions are closer to individuals than international institutions, the growing interdependence between States has led to the expanding reach of international law.

Individuals are directly affected by international law and policy, for example in conflict situations, humanitarian interventions and through increasing international trade rules, and now through the attempts to create governance mechanisms in bioethics and the human genome. Increasingly the mechanisms we have in place are surpassing State borders and if this is considered as a trend for the future, then the international collection of States can be looked at as one large federation, one large State. In Gewirthian theory, States are envisaged as being 'supportive', in whose territory a 'community of rights' prevails. The need for a supportive State, which could otherwise be substituted by the concept of the minimal State,[18] arises under the following conditions that might prevail in the State:

> Many persons lack adequate food, housing, necessary medical care, and other basic goods, and they also lack the capacities to assure that they will continue to have these goods as needed; they may be subjected to degrading conditions of dependence, danger and disease in their situations of life and work; they may also not have the means to increase their capabilities of purpose – fulfilling action, such as adequate education and income.

Within this hypothetical large federation, disabling and disempowering circumstances prevail, both within individual nation-states and unfortunately, as a characteristic of a nation-state – for instance, the least-developed countries.[19] A case for a supportive international community working towards the fulfilment of the basic goods required to sustain human action can be constructed.

If rules already exist in international law and are applicable to the international community, the question is whether the current exercise is only to show that they are good and that they should continue to prevail. Partially, in examining the notions of the common heritage of mankind (CHM) and the RTD, my aim is to clarify the ethical and moral basis of such notions and where present, to bring out explicitly their ethical and moral basis, but it is also the purpose of this chapter to argue for a wider recognition for these concepts for the governance of biotechnology and health.[20]

Gewirth's 'community of rights' refers to the embodiment of a notion of entitlement within economic and social policies aimed at relieving human suffering.[21] Suffering should be understood in the above-mentioned context as related to the deprivation of basic goods. Within the community of rights, Gewirth picks out a few rights as being important because of the strategic role they play in the fulfilment of other rights. He focuses on five rights, namely, the right to productive agency, the right to private property, the right to employment, the right to economic democracy and the right to political democracy. In doing so, he emphasises the need for political systems and philosophies to be also based on empirical understanding of economic and social problems.[22] These rights are primarily related to economic and social rights within the framework of international human rights law but reinforce the indivisibility and interrelatedness of human rights[23] through the importance placed in Gewirthian theory on the notion of 'freedom'. Gewirth argues for procedural freedom and substantive rights as part of the fulfilment of needs required for human action.

The RTD and the CHM framework reinforce the indivisibility of human rights but also primarily approach rights from the economic and social point of view. The argument is for structural changes in the international economic system in order to respond to the fulfilment of economic and social rights, while at the same time empowering individuals and communities to enjoy and support civil and political rights in a meaningful way, which is very similar to the Gewirthian project of governance. In Gewirthian theory there is a clear hierarchy of rights and onerous obligations entail basic rights. The notion of hierarchy is prevalent in international law. The International Court of Justice (ICJ), in the Barcelona Traction case, also created a hierarchy in rights when it alluded to fundamental human rights protection against slavery and racial discrimination.[24] The notion of hierarchy in rights is an important one in the context of the governance of the human genome and development issues – specifically in relation to prioritisation in research and of resource use.

After this brief introduction to Gewirthian theory, the structure of this chapter will now be summarised. It sketches out Gewirth's conceptions of generic needs, generic rights and the PGC, which links these notions by applying rational rules and spells out a theory where agents are mutually dependent on each other for the fulfilment of basic needs. A dialectically 'necessary' method is then used to show that agents should accept the PGC – otherwise they contradict their status as agents. This is followed by the use of a dialectically contingent method, coming from the position that a notion of human rights exists, which is in turn followed by a dialectically contingent argument which leads to the conclusion that those who believe in the existence of human rights should also accept the PGC if they do not want to contradict themselves. The oppositions and criticisms[25] to the principle of PGC are not fully dealt with here, as the author considers most of them to have been addressed satisfactorily by other commentators.[26]

A departure is made from a discussion of Gewirthian theory and its significance is examined before a further analysis of Gewirthian concepts such as positive rights, mutuality, the reasonable self, the idea of community, and ethical individualism is undertaken. With this the stage is set for the examining the elements of a CHM regime through the lens of Gewirthian theory.

Gewirth sets out to make a strong case for economic and social rights,[27] and since his is a hierarchical ordering of rights, he delineates and elaborates a notion of the RTD. Gewirth's notion of the RTD is juxtaposed alongside the notion already examined in the earlier chapters in this volume and a rational human rights-based definition of the RTD is developed. With respect to the concept of the CHM, it is concluded that it is designed to fulfil basic needs, especially through its emphasis on the mutuality of agents and the resultant governance structures through international cooperation mechanisms, which are taken to reflect mutuality.

5.3.1 Generic needs of agency (GNA) explained

Using pure logic as the criteria for rationality, Gewirth derives a set of needs for an agent, which he qualifies with the term 'generic', that are fundamental to the agent itself. If the agent denies generic needs, then it contradicts itself. An agent, by definition, is a being (not only humans) that can voluntarily pursue actions for a self-chosen purpose or have an inherent capacity (simultaneous with an inclination to do so) to voluntarily pursue actions for a self-chosen purpose.[28] All agents act. (All agents value their self-chosen purpose and therefore value its pursuit.) To act in pursuit of its purpose, at the very minimum, the agent needs to be *able* to act. Therefore even before action, the agent needs to be endowed with the *ability* to act.

The prerequisites of an ability to act at all or with[29] are identified as the 'generic needs of agency' (GNA). Gewirth's approach to philosophy is foundational, in that it is based on a generic practical truth that all agents act for certain purposes.[30] At the same time, there is no claim made that the theory is the truth.[31] There is no claim that all agents are rational. The method that Gewirth uses to derive the GNA is a dialectic one. It is as if someone were thinking aloud – it is a form of a dialogue (but one that is carried out with oneself) – to arrive at a truth. Gewirth employs a dialectically 'necessary' method, implying reliance only upon the barest of truths to start with. Such a method makes it more compelling to accept the final derivations that follow from it than if they were to be argued from a dialectically contingent position, whereby the initial premises are true only under certain conditions. In fact, it will be argued later, contingently, that any international order which purports to grant human rights should recognise that the basis of human rights are the GNA (else it contradicts itself). Although a dialectally contingent argument for generic needs is not as foundational as a dialectically necessary agreement, the foundation of human rights is a strong one[32], therefore a contingent argument from the premise that human rights exist is still a persuasive argument.

An argument from a dialectical standpoint of the agent would be:

I am an Agent – I act. For me to act I need to be *able* to act since I value the pursuit of this act. Therefore I need the minimum means to act. These minimum needs are my generic needs to agency.

To give flesh to the concept of the GNA in case of the human species, food[33], clothing, shelter and health[34] can be considered to fall into the category of GNA.[35] At a more fundamental level, GNA include life itself, the capacities involved in an ability to make choices, the possession of mental equilibrium sufficient to enable one to translate one's preferences into active pursuit of one's purposes, as well as the necessary means to these (meaning those things the absence of – or interference with – will threaten or interfere with life, etc.).[36] There are two other categories of needs which are not generic. One can just be able to act; one can act successfully and one can achieve great success in one's actions. Corresponding to these three levels of actions are the three different tiers of needs, the first of which is the GNA. The GNA are also called basic needs, followed by non-subtractive needs and additive needs respectively. Gewirth uses the term 'goods' instead of 'needs', from the angle of the facilitator of agency who would secure to the agent these goods which correspond to the needs of the agent.

5.3.2 From generic needs to generic rights

Because the agent accepts that GNA are essential for performing any action at all, she must also accept that she has an interest in having GNA, since it is the instrument through which she can act at all. She has no other alternative but to possess GNA (or contradict herself as an agent – which is not possible at this point since she has accepted GNA). In this, she must also recognise that others should not hinder her from obtaining her GNA (negative claim rights) and should instead help her obtain her GNA when she cannot do so herself. Therefore she has both positive and negative claim rights to GNA. GNA give rise to generic rights of agency. The conception of rights in the Gewirthian theory is not only negative – because the GNA are vital to being an agent positive claim rights to GNA are also integral to the analysis.

This is the most notable feature of generic rights. International cooperation, solidarity and dignity depend on such a formulation of rights. Since GNA are categorically instrumental for action, they have to be necessarily secured to the agents for generally successful action and, in cases where agents cannot obtain it themselves, by others. The others have a duty towards agents to positively help them secure their GNA. As Kant puts it 'rational beings, with needs, united by nature in one dwelling place for the purpose of helping one another have a universal duty of beneficence toward the needy.'[37]

Generic rights are claim rights since others have a correlative duty towards the holders of these rights. There are two types of conception of rights – will conception and the interest conception. According to an interest conception of rights, one possesses certain rights because it is in the interest of such a possessor to possess these rights; they cannot waive the exercise of such rights. For example, under the interest conception of rights, one does not have a right to commit suicide since it is against one's interest not to stay alive. Whereas under a will conception of rights, one can exercise one's will to waive the rights one holds.[38] Four main aspects of generic rights need highlighting. They are:

- strong or claim rights[39]
- positive as well as negative

- rights under the 'will conception' of rights
- moral rights.

Gewirth defines morality as 'a set of categorically obligatory requirements or action that are addressed at least in part to every actual or prospective agent, and that are concerned with furthering the interests, especially the most important interests, of persons or recipients other than or in addition to the agent or the speaker'[40]. The most important interests refer to the GNA. If agents are not able to secure to themselves their GNA, then others have to help them secure it. In the same vein, if others cannot secure their GNA, agents have to rush to their help (within the capacities of what they can do) to help them secure their GNA. This is an obligatory requirement, otherwise the agents deny that they possess GNA. Therefore generic rights are also moral rights, since they are obligatory, other-regarding rights[41].

5.3.3 The Principle of Generic Consistency (PGC)

Having started with GNA and looked at its link to generic rights, the next step is to see how the PGC is established from the premise that generic rights entail GNA. For the sake of continuity, I will restate the need for GNA and generic rights and then proceed to establish the PGC.[42] The dialectically necessary method is employed.

By claiming to be an agent, I claim (by definition):

(1) I do (or intend to do) X voluntarily for a purpose E that I have chosen.

Because E is my freely chosen purpose, I must accept:

(2) E is good, meaning only that I attach sufficient value to E to motivate me to pursue E (i.e. that I value E proactively). If I do not accept (2), then I deny that I am an agent – which is to say that it is *dialectically necessary* for me to accept (2).

(3) There are generic goods of agency.

Therefore, I must accept:

(4) My having the generic needs is good *for* my achieving E *whatever* E *might be* ≡ My having the generic needs is categorically instrumentally good for me.

Because I value my purposes proactively, this is equivalent to my having to accept

(5) I categorically instrumentally ought to pursue my having the generic needs.

Because my having the generic needs is necessary for me to pursue my having the generic needs, I must hold:

(6) Other agents categorically ought not to interfere with my having the generic needs against *my will*, and ought to aid me secure the generic needs when I cannot do so by my own unaided efforts *if I* so wish, which is to say, I must hold:

(7) I have both negative and positive claim rights to have the generic needs ≡ I have the generic rights.

It follows (purely logically) that I must hold, not only (7), but also:

(7') I am an agent → I have generic rights.

Consequently, it follows (purely logically) that I must hold:

(8) All agents have the generic rights.

Since I deny that I am an agent by denying (8), every agent denies that it is an agent by denying (8). Thus, (8) is dialectically necessary for every agent. To put it in simpler language, specific to human agency, 'A person is generically inconsistent, or inconsistent in a fundamental way, if, while exercising his own capacities of agency, he rejects, either in thought or in action, the possession or exercise of those capacities on the part of other persons'.[43] This statement reflects the PGC, which is stated in (8). All agents, including the agent who dialectically argues for the PGC, have generic rights. The establishment of the PGC from the GNA-generic rights derivation further broadens the scope of the duties arising from positive rights that fall upon an agent. Therefore, from a position that one is an agent, a purely logical analysis leads one to conclude that agents have to help secure the generic rights of agents.

5.4 The PGC as the basis of the international human rights order

The PGC or the recognition of the generic needs of all agents is the basis of the international human rights order as elaborated by the 1948 Universal Declaration of Human Rights and other founding instruments. Generic rights are also human rights. The argument to establish the PGC from the premise that GNA are implicit in agency is also the argument for the existence of moral rights of agents or. in other words, human rights. Human rights are widely understood as the rights endowed in a human being by virtue of being a human being.[44] Human rights are also widely held to be rights that are, by their very nature, universal and are independent of a positive recognition of these rights since they are based on natural law principles. Hersh Lauterpacht traces the origins of human rights and freedoms to natural law and natural rights, which underlies them as constituting the 'higher law' which must forever remain the ultimate standard of fitness of all positive law, whether national or international.[45] However, a rational argument for claiming that human rights are universal and must be granted by virtue of the 'humanity' in the human being is not always clearly enunciated. In the following paragraphs, the PGC is offered as a justifying principle for the existence of human rights.

In the PGC, when the agents in question are identified as humans the generic rights that belong to these agents are nothing other than human rights. The PGC then becomes the essential criterion of legal validity for the existence of human rights. Gewirth sets out to show in his essay entitled 'The Basis and Content of Human Rights'[46] that all human rights have a rational foundation in the necessary conditions or needs of human action or GNA, so that no human agent can deny or violate them except on pain of self-contradiction. He goes on to say that the demands human rights make on persons are justified by the PGC as the supreme principle of morality. Also, through his dynamic-instrumental justification of social rules, he urges institutions and governments to promote human rights in the same way that agents are bound to promote the human rights of other agents.

The question arises as to whether all agency needs, which include the basic, non-subtractive and additive goods, are to be considered as part of the human rights framework that should be enforced by law. Gewirth raises this question and answers it thus:

> Not all the human rights upheld by the PGC should receive legal enforcement. The specific harms done by violations of a person's non-subtractive rights, such as when he is lied to or when a promise made to him is broken, are ordinarily less important in their impact on their recipient's well-being than are the harms done by violations of basic rights, and hence do not justify the state's coercive legal resources to combat or correct them.

Therefore human rights laws should concentrate on securing to the human agents their fundamental generic rights or the basic rights, on a priority basis.

Beyleveld and Brownsword have argued that it follows from the acceptance of human rights that legal systems that recognise human rights must treat the PGC as a necessary criterion of legal validity, on pain of denying that they recognise human rights.[47] They arrive at this conclusion based on a dialectically contingent argument for the PGC from the acceptance of human rights. The argument has been presented in two phases. The first one endeavours to show that anyone who accepts that there are human rights has to accept the GNA. The second phase delves into the will conception of rights and justifies the PGC on the basis that the international human rights order is actually based on a will conception of human rights, and not on an interest conception of human rights.

Phase 1: A dialectically contingent argument for GNA based on human rights

Suppose I (any agent) claim:

(1) I have a right to do X.

I must also claim:

(2) I have a right to the necessary means to do X.

Since:

(3) There are generic needs of agency.

I must also claim:

(4) *Whatever X* is, I have rights to the generic needs of agency.

From this, it follows logically that:

(5) Any agent who considers that it has a right to do anything must consider that it has rights to the generic needs of agency.

Hence, it also follows that:

(6) Any agent who considers that it has a right of kind R to do anything must consider that it has rights of kind R to the generic needs of agency.

Thus, it follows that:

(7) Any agent who considers that it has a human right to do anything must consider that it has human rights to the generic needs of agency.

Considering that human rights are purported to vest in human beings by virtue of being 'human', it follows that:

(8) Any agent who considers that any being has a human right to do anything must consider that the being has rights to the generic needs of agency simply because it is human.

From this, it follows that:

(9) Any agent who considers that any being has a human right to do anything must consider that all humans have rights to the generic needs of agency.

Phase 2: Justification of the PGC based on an idea of human rights founded on the will conception of rights

Under a will conception of rights, the holders of the rights can waive the benefits that arise from the exercise of these rights. In this case, it is the generic rights. Beyleveld and Brownsword argue that 'the fact that the human rights instruments do not outlaw boxing, mountain climbing, and other dangerous activities, and do not require persons to vote etc., shows that the will conception (which is only applicable to agents) is to be applied to at least some of the rights'.[48] They argue that the extension of human rights to certain special groups like foetuses, children, the dead etc. are fairly recent and controversial. According to the PGC, the rights granted to the agents will be commensurate to their ability to exercise their agency. So it is possible to be a partial agent. Since the majority of human rights granted under different legal regimes conform to a will conception of rights, in the interests on coherence, it should be taken that international human rights order works under the will conception. In cases where courts have interpreted under an interest conception, it should be treated as a mistake. Therefore, from step (9) above it follows that:

(10) Any agent who considers that any being has a human right to do anything must accept the PGC.

One of the significant conclusions that one reaches from a dialectally contingent argument for the PGC is that human rights instruments like the International Covenant on the Economic, Social, and Cultural Rights, the International Covenant on Civil and Political Rights, etc. must, on the pain of contradicting that it is in fact a convention on human rights, be interpreted in line with the requirements of the PGC. In the same vein, the PGC can be used to supply clarifications on human rights instruments when there are doubts as to its meaning. Where a right has not been explicitly granted, the PGC can supply it. And when, with the progress of technology, new developments are sought to be regulated, the PGC can be used as a legislative guide to enunciate the different rights and restrictions that can be placed on the technology.

The 1997 Universal Declaration on the Human Genome and Human Rights is one such instrument that has been necessitated by the march of biotechnology in general and genomics in particular. It does not have binding force in law, although it could form the basis of customary law formation in the area of genomics and human rights.[49] The Universal Declaration is the first instrument to deal with human rights issues in the human genome. It broadly covers areas such as privacy and data protection of the research subject, rights and duties of scientists, and the equitableness of genome research. It has been shown that the Universal Declaration supports a CHM regime, which has at its core the RTD. A Gewirthian analysis of the RTD will be undertaken in the following sections.

5.5 Some important concepts in a Gewirthian analysis

5.5.1 Positive rights

As mentioned earlier, generic rights are positive as well as negative rights. Positive rights are rights that are vested in persons and which come into play when persons cannot secure their freedom and wellbeing by themselves. By virtue of having positive rights they have a right to be helped by other persons and, if need be, institutions, to secure these rights. This is an important aspect of the PGC. In fact, Gewirth establishes a 'community of rights' primarily based on the recognition of positive rights. He emphasises that 'when such help is needed by large numbers of persons, and especially when their needs have institutional roots, such help requires a context of institutional rules, including the supportive state as the community of rights'.[50] The notion of positive rights is controversial in international human rights law. The scepticism about positive rights is clearly expressed in the following extract:

> the traditional political and civil rights are not difficult to institute. For the most part, they require governments and other people generally, to leave a man alone .... Do not injure, arrest or imprison him .... Since those rights are largely rights against government interference, the greatest effort will be directed toward restraining the government's own executive arm. But this is no longer the case with economic and social rights.[51]

The grounding of rights in human needs, as is done by Gewirth, dispels some of the controversies surrounding positive rights and provides a basis for why a human rights system based on morality and rationality should treat positive rights on par with negative rights.

A case has been made for a difference in treatment of positive and negative rights, based on the 'ease' of protection of these rights. Although it might sound like a pragmatic approach to human rights, it is not morally defensible, as already noted earlier in this book.

5.5.2 Mutuality

The difference between mutuality and reciprocity is the subject of this section. Positive rights that are central to the community of rights are mutual in nature, and not reciprocal. Gewirth deals at length with the difference between mutuality and reciprocity and locates positive rights not so much in reciprocity but in mutuality, by recognising 'a common status of having needs that require for their fulfilment the positive help of others'. While reciprocity can be understood as tit for tat – equal amounts of something exchanged between two persons – mutuality should be understood as giving or receiving something so as to secure to the other his or her generic rights. In a mutual relationship, within the international community, for example, individuals and institutions would work to secure the generic rights of persons if they are not capable of securing them for themselves. The Millennium Development Goals (MDGs) are an illustration of the 'mutual' commitment by the international community to secure to the world's poor their generic rights because of their incapacity to secure it for themselves.[52] Because of increased cooperation between States and a rapidly globalising world, the State in some cases has now been supplemented/displaced by the international community.

The other aspects that distinguish mutuality from reciprocity are:

- In a mutual relationship this giving or receiving can be simultaneous. But, in a reciprocal one, B gives something to A only if A has given something to B.
- In a relation of reciprocity the beneficiaries are limited to one's prior benefactors, unlike in mutuality, where they are owed to all agents.
- Another difference between reciprocity and mutuality is temporal. 'The antecedent of the reciprocity relation is a contingent one: if A has received some good X from B (this receipt being contingent on an optional decision by B) then A owes a fitting response to B' but 'in the mutuality of human rights, ... the antecedent is itself necessary: because other persons have human rights against you (these rights not being contingent for their generation or validity on some prior optional beneficent actions on their part), it necessarily follows that you have correlative duties to them'.[53]

Positive rights, based on mutuality, are reaffirmed in the 1997 Universal Declaration. Its provisions on information and benefit sharing seem to reflect a Gewirthian community of rights perspective.

5.5.3 Reasonable self and ethical individualism

For a community of rights to function effectively, Gewirth provides further justification of the close relations that the community and its members have through the 'social contribution thesis'.[54] He urges that there is a rational basis for recognising society's contribution to successful purposive agency and vice versa. Since, in a community of rights based on a recognition of positive rights, the society, through its individuals and institutions, contributes to the fulfilment of the agents' generic rights, the agent and all other agents who are members of the community are 'social products'. The nurturing they receive from their social context, beginning with the family but continuing into the wider society, is a necessary condition of their being successful purposive agents.[55] In recognition of the contribution of the community to their agency, the members have to accept an obligation to the society. Such an agent then behaves as a 'reasonable self'. A reasonable self is aware of its obligations towards others as well as rights against others, and that these must be embodied not only in individual actions but in social and political institutions, whereby they constitute a community of rights.[56]

A reasonable self also has a sense of personal responsibility towards one's own agency needs and rights since its first and foremost duty is to try to fulfil one's agency needs, failing which, the 'positive' nature of generic rights comes into action, and others have a duty to help secure the generic rights of the agent acting as a reasonable self. A reasonable self is aware of its own agency needs and rights and it has a sense of personal responsibility for their fulfilment. It also takes due account of the agency needs of other persons, respecting their rights as well as one's own and maintaining a certain equitableness or mutuality of consideration between oneself and others, as required by the universality of human rights. Apart from a sense of personal responsibility, the reasonable self also acts with autonomy. Autonomy of action is subject to mutualist considerations. Autonomy is different from freedom because it includes the idea of setting a rule or law for oneself – it requires that one:

> exercise control not only over particular actions or behavioural episodes but also over the general pattern of one's behaviour, including the ends of one's actions as well as their means. Autonomy requires some degree of awareness of this general rule, including one's ends, as being adopted or maintained by and for oneself. Thus autonomy is not only occurrent but also dispositional; it involves long-range, principled, reflective control over one's behaviour, including both what one does and what happens to one. But, as with freedom, this control must derive from one's informed and unforced choice as the sufficient condition of one's behaviour.

Exercise of such autonomy, when laced with mutualist considerations results in 'rational autonomy', in that 'the rules or laws that one sets for oneself have been arrived at by, or at least compatible with, a correct use of reason so that one recognises that all other prospective agents have the same rights one necessarily ascribes to oneself'.[57]

A reasonable self, acting in recognition of the positive rights of other agents in a society that reflects a community of rights, then, comes from a point of ethical individualism. Ethical individualism is 'a social relation bearing on the ends of social policies and institutions'. It is characterised not by the egoistic drives of some individuals against others but rather by a whole system of equal and mutually supportive rights and duties, presided over by institutions, which are the constituent parts of this system. These institutions lay down rules, which require systematic contributions by all persons, as far as possible, to maintain this system of society. A reasonable self, operating from an ethical individualist standpoint, fully understands the need for these institutions and rules and cooperates with them.

5.5.4 The right to productive agency

Gewirth traces the right to productive agency as one of the main economic and social rights that effectuates the principle that each person should be equally protected in having the freedom and wellbeing that are the proximate necessary conditions of action and generally successful action (the PGC as the basis of human rights). Of the two areas of rights, social and economic, Gewirth considers the latter to be significant since it also helps secure the former, in many ways. He justifies the choice of his focus on economic rights (which is the argument that this book relies upon too, since it primarily analyses the economic aspects of the 1997 Universal Declaration on the Human Genome and Human Rights) because 'they are strategically important for the problem of attaining for all persons, the freedom and well-being that are the necessary conditions of human action and generally successful action'. He also goes on to say that 'the incurring of disease and the availability of health care are in themselves strongly influenced by economic factors'[58], words which are also relevant to the subject matter of this study, namely genomics and health.

Using the 'deprivation focus' as a starting point, i.e. focusing on how to effectuate the generic rights of the most deprived, Gewirth derives the right to productive agency as the ability to be secure to oneself, to have 'additive wellbeing'. It is appropriate here to elaborate further on what additive wellbeing is and how it relates to the GNA. Gewirth classifies the GNA into freedom and wellbeing. Wellbeing itself is of three kinds: basic, non-subtractive and additive. Beyleveld and Brownsword group the three kinds into two categories: things needed for the very possibility of action, which include the basic goods and things needed for the possibility of successful action, which include the non-subtractive and additive goods. To illustrate the usefulness of this classification, one only needs to imagine a conflict of generic rights that could easily arise in any situation. In case of conflict, 'basic rights will trump non-subtractive ones, which will trump additive ones, and finer distinctions within these qualitative categories will be possible quantitatively'.[59]

Therefore, the right to productive agency, which in itself should be perceived by an agent as a necessary good for herself, is the right to one's basic wellbeing, which, most significantly, includes also the right of the agent to have the ability to secure to herself non-subtractive and additive wellbeing. Through such an enunciation of the right to one's basic wellbeing Gewirth has been able to achieve what many other theories lack – the autonomy of the agent. While reminding oneself that the generic rights are positive rights, the right to productive agency bears upon a productive agent as a duty to help other agents secure their generic needs, which include the right to the ability to secure for themselves their non-subtractive and additive needs. They will not be continuous passive recipients of help from other agents, thereby acquiring a dignified life.

The right to productive agency is a therefore a two-fold right – the right to the fulfilment of basic needs of agency, i.e. the right to the possibility of any action at all, and the right to be able to act with any possibility of successful action, as and when the agent undertakes such action. Hence 'all persons have positive rights to the development of such general productive agency as part of their generic rights to both freedom and well-being'.[60] Gewirth also highlights that most people, as an empirical fact, have to make an income or acquire some wealth, through their own work, sufficient enough to protect their freedom and wellbeing.

From the angle of the institutions and the way in which a community of rights should be structured to allow the right of productive agency to flourish, this is facilitated by the ethos of 'productivist welfarism'. In a 'productivist welfarist' community the focus is on distributing and fostering personal productive abilities or resources and not directly on distributing products and money, as has been the case with traditional theory which is supported by international development aid, etc. based on dependence and passive welfarism. Such a productivist welfarist community would then fulfil the ultimate purpose of generic rights and of human rights, which is to 'secure to each person a certain fundamental moral status: that each person have rational autonomy in the sense of being a self-controlling, self-developing agent, who can relate to other persons on a basis of mutual respect and co-operation, in contrast to being a passive, dependent, recipient of the agency of others'.[61] Thus a welfare State, or the international community aiming at such a position, becomes a conundrum, in the sense that its purpose is to ensure that it is constantly moving towards the condition where it no longer has to be a welfare State.

A reasonable self who is a productive agent, and is a part of such a productivist welfarist community, would support and contribute to such a community and go along with its schemes, as well as actively participate in schemes that are targeted at fostering the right to productive agency.

5.5.5 The right to private property

Gewirth argues that in so far as there is a right to productive agency, there is a right to private property, at least as represented by one's earned income and accrued wealth.[62] It is also a positive right, just like the right to productive agency. There are two justifications to this right: the consequentialist justification and the antecedentalist justification. The consequentialist justification bears on the final cause, or the end purpose, served by these rights to the holders of these rights. Having a right to private property is essential to fulfil GNA, therefore such a right is justified. One of the antecedentalist justifications of private property rights is the purposive labour thesis, whereby a productive agent has the right to property which he has produced through the exercise of his agency. One's labour should be rewarded by having rights in what it has helped to produce. According to Beyleveld and Brownsword, the consequentialist argument should be treated as the primary justification for private property rights, and this author concurs with this statement. Therefore, 'in the absence of conflicting claims on goods and services by others, persons must be granted property control over objects that satisfy the needs of a consequentialist justification, and consequentialist considerations would be sufficient'.[63] But because conflicting claims are always the reality, Beyleveld and Brownsword state that the antecedentalist claim to private property should always be subservient to the consequentialist claim because such property rights 'can never operate validly so as to deprive A of the generic needs of agency when there is sufficient of X to satisfy A's basic needs without depriving B of B's basic needs'.[64]

Beyleveld and Brownsword define property rights as secondary rights, which are required by primary rights. The function of claims to property rights can be said to be 'preclusionary', according to Beyleveld and Brownsword. Because, as a rule, it is essential to have property rights, in order to be able to protect the GNA, they call their theory of property, the rule-preclusionary theory of property. Property rights are no more than a prima facie entitlement to exclude others,[65] and that which require no other justification other than that they are required for the fulfilment of the agent's needs. In elaborating the rule-preclusionary principle, Beyleveld and Brownsword pose the question 'Why should it be recognised that persons (generally or individually) have property in P?', and answer it with 'Because not granting persons generally (or that specific person) preclusionary control over P, as a *general rule*, does not provide adequate protection for their specific rights'.[66]

5.6 A moral analysis of the notion of the CHM

The ethos of the CHM regime is that of the productivist welfarist society that Gewirth envisaged and the role of individuals within such a regime is based on a conception of ethical individualism. The elements of the CHM correspond to themes and ideas in the Gewirthian moral framework in the following way:

1 Non-appropriation – negative right.
2 International governance – positive right to assert freedom by taking part in decision making.
3 Peaceful uses – protection of basic human rights, such as the right to life.
4 Benefit sharing – development of productive agency, a duty to cooperate (reciprocity).

5.6.1 Non-appropriation as a negative right

Non-appropriation in the context of the human genome and the seabed was formulated for the purpose of retaining open access. Appropriation would entail the loss of control over the resources in the seabed resulting (possibly) in non-sharing in the resources and lack of decision-making power with respect to the Area. In order to to enable the retention of non-exclusive control over the seabed, the non-appropriation clause was formulated. According to Article 4 of the 1997 Universal Declaration on the Human Genome and Human Rights, 'The human genome in its natural state shall not give rise to financial gains.' In the context of a CHM regime the clause indicates that the information in the human genome shall be free and open to all.[67] The policy for the publicly funded Human Genome Project (HGP) was to put the results of the sequencing into the public domain so they could be used as raw data, which is essentially the first step towards progress in putting the 'genome to use' in order to derive benefits from it. The Bermuda Rules, first agreed upon in 1996, collectively are a demonstration of the principle embodied in Article 4 of the 1997 Universal Declaration. Under the Bermuda Rules, the information on the sequenced gene should be available within reasonable periods of time for others working in the same field, with as much accuracy as possible.[68] It stipulates that such information should not be taken advantage of by other scientists and that appropriate citations should be made of the original sequencers.

The non-appropriation clause in the 1997 Universal Declaration confers a negative right on individuals by facilitating the free access to information essential to the realisation of the basic needs of agency, amongst which is health. It leaves the information unencumbered by proprietary considerations and therefore paves the way for its further use.[69] It is similar to the 1982 Convention on the Law of the Sea and the 1979 Moon Treaty whereby the seabed is prevented from being appropriated as sovereign territory since such an arrangement would hinder any further involvement by parties other than those who have control and authority over that area. In the Moon Treaty, although the actual CHM regime was put on hold till further developments in moon exploration revealed the existence of useful resources, the non-appropriation clause was put in place so as to maintain the freedom of access that is crucial for the elaboration of future regimes.

5.6.2 International governance as a positive right

International governance is a mechanism through which the positive right of individuals to exercise their freedom, by taking part in the decision-making process, can be exercised by bringing all actors to cooperate on a particular theme.[70] The formal equality of States is maintained in the CHM regime and the right to equal participation is given effect to, in the regime. The interests of disadvantaged States, such as landlocked and island States, are given special attention as are also the developing countries and the pioneer investors. The needs of the countries and their economies are taken into consideration in such a system of governance. This in turn, is designed to have the effect of creating favourable circumstances for the fulfilment of the RTD. International governance is essential to counter the power of entities, including States which can claim a larger role in decision making and curtail the freedom of States with lesser power; hence it is essential to establish good governance practices, with an effective voting system which would give effect to the equality of States.

5.6.3 Peaceful use – the right to life

Peaceful use is an essential condition of a CHM regime, as non-peaceful uses can have an adverse effect on the fulfilment and protection of human rights. The installation of nuclear facilities or military facilities on the seabed or the moon would have disastrous consequences for the right to life and the basic needs of human agency. In the case of the human genome, non-peaceful uses such as creation of dangerous mutant pathogens using genetic techniques or chimeras which are violent would lead to gross violations of the right to life.

5.6.4 Benefit sharing: reciprocity, development of productive agency

Benefit sharing plays an important role in the fulfilment of the positive right to the development of productive agency. In the context of the human genome, health is an important basic good that is essential for the fulfilment of basic needs of human agency.[71] The benefit-sharing clauses in the 1997 Universal Declaration aim at fulfilling the basic needs of human agency to those who are currently not in possession of it. This can be illustrated through Article 12 which stipulates that the 'applications of research, including applications in biology, genetics and medicine, concerning the human genome, shall seek to offer relief from suffering and improve the health of individuals and humankind as a whole'.[72] This provision is supplemented by another provision which States that 'research on the human genome and the resulting applications open up vast prospects for progress in improving the health of individuals and of humankind as a whole'. At another level, these provisions also aim at closing the gap between the levels of health enjoyed by the poor and the rich and to bring about equality.

The commitment to basic needs is visible in the use of the phrase 'relief and suffering' while the commitment to enhanced health, over and above the basic needs of agency (moving towards the fulfilment of additive and non-subtractive rights), also forms part of the 1997 Universal Declaration. Solidarity and cooperation are seen as crucial to the benefit-sharing project. Within a framework of solidarity, the exchange of expertise, knowledge and technology is encouraged.[73] The community for which 1997 Universal Declaration has been drafted can be easily situated within the 'deprivation focus' that Gewirth uses as his starting point to discuss the right to productive agency and the State as a productivist welfarist State. The international community is comprised of a majority of people who are deprived of their basic needs and unable to live life in dignity. The notion of solidarity within such a community creates favourable conditions for the fulfilment of the mutuality of rights, which is a characteristic of human rights.

In the event that countries are unable to effect improvements in the health of the individuals that populate them it falls upon other countries, which are envisaged as communities of rights, to help and aid them by cooperating with them in order to protect the human rights of their citizens. The exchange, in some cases, may be a one-way street, with no quid pro quo involved, both in the case of fulfilling basic needs and as a project to bring about social justice.[74] The sense in which social justice is used here is broad and meant to encompass various charitable activities of groups and States in the belief that such actions are just and somehow owed to the suffering peoples. Although not based on an idea of mutuality of agents, such an idea is related to mutuality because it recognises positive rights. The emphasis on capacity building and technology transfer corresponds to the need for autonomy and freedom. Not only is it important that the health of individuals be improved or maintained to the extent that the basic needs of agency are fulfilled, it is essential that they also acquire the ability to continue to remain in this state, through their own capacities, and also by way of support from institutions that they have created, such as States and governments.

5.7 The RTD and productive agency

The RTD is a fundamental human right which underpins the CHM regime. In the Convention on the Law of the Sea it gave rise to a regime which allows for equitable sharing of natural resources on the seabed to aid the struggling economies of poor countries in order that their peoples enjoy social and economic progress. The Moon Treaty also adopted the CHM principle in order to delineate an equitable regime. The era of the new international order gave way to a less ambitious agenda whereby developing countries adopted a human rights approach to development, amongst others – the opening up of their economies to global trade, participating in global regulation of trade and commerce and asserting their sovereignty over natural resources were other prominent approaches adopted by developing countries in order to improve their economic situation and enhance their status in the international community. The human rights approach was used to establish the RTD.

The RTD is a minimalist approach to development which seeks to fulfil the right to participate in, and contribute to, the process of development. It is also a right to a process, which empowered individuals to achieve certain basic capabilities. Amartya Sen calls it a 'meta right' through which all other rights can be realised.[75] Norman Daniels talks about the goal of enabling people to utilise a range of opportunities available to them.[76] Henry Shue calls it a basic right.[77] Gewirth identifies the right to productive agency as 'the agent's ability to achieve the outcome he desires or intends', adding that 'a productive agent is one who does thus achieve'[78]. These ideas have similarities both to each other and to the notion of RTD.

Gewirth claims that 'all persons have positive rights to the development of such general productive agency as part of their generic rights to both freedom and well-being.'[79] Freedom is the procedural aspect of the right to productive agency, and wellbeing is the substantive aspect of the right to productive agency. Both are essential in fulfilling the right to productive agency. The 1986 Declaration on the Right to Development views the RTD as an aggregate of civil, political, social, economic and cultural rights[80] and the right to productive agency, by embodying the generic rights to freedom and wellbeing, closely resembles the RTD. Gewirth explains the notion of productive agency with reference to his classification of wellbeing, which is hierarchical and includes basic, non-subtractive and additive wellbeing. He states that 'productive' is synonymous with 'additiveness', as used in 'additive well-being'.[81] This notion of productivity corresponds to the 1986 Declaration which deals with the RTD as the ability to 'contribute to' development. Gewirth uses the example of a weightlifter who is deemed to be a productive agent if he is successful in lifting his weight. He is capable of achieving his self-chosen purpose, which is to lift the weight. The acquisition of such capability is tantamount to the fulfilment of productive agency. This corresponds to the RTD as a right to acquire the capability to be able to 'participate in' development, which is the other essential criterion in the RTD, as laid out in the 1986 Declaration.

Gewirth also elaborates a process that enables the fulfilment of productive agency, which he calls productivist welfarism. Health and health care are fundamental to the fulfilment of the right to productive agency, as Gewirth himself indicates,[82] and are also fundamental to the RTD, as it is stated in Article 8 of the 1986 Declaration that States shall ensure 'all necessary measures for the realization of the right to development and shall ensure, inter alia, equality of opportunity for all in their access to basic resources, education, *health services*, food, housing, employment and the fair distribution of income' (emphasis added).[83] Identifying 'the vast area of illness and required health care as a prominent area' from which to base his notion of deprivation focus, from which in turn he derives the right to productive agency, Gewirth states that:

> there are many other economically based kinds of affliction besides the ones that will be discussed here [in his book *The Community of Rights*] .... My justification for focussing on the areas I discuss here is that they are strategically important for the problem of attaining for all persons the freedom and well-being that are the necessary conditions of human action and generally successful action.

Gewirth makes an important point about the link between economic factors and the availability of health care.[84] Underpinned by the RTD, the CHM regime, with its emphasis on the economic factors that affect development, will serve to maintain the link between health and economic factors and, where need be, create and reinforce such a link.[85]

The CHM framework provides scope for consideration of health, human rights, international trade and intellectual property rights as part of the project that contributes to the fulfilment of the basic needs of human agency. It has been a feature of human rights that civil and political rights are given emphasis over economic, social and cultural rights. International trade has been concerned more with opening up new markets and reducing barriers to trade than it has with development and intellectual property has been about enhancing protection for the rights of the inventor and authors without paying much attention to the benefit it might bring to society. The CHM regime for the human genome envisages a new kind of productivist welfarist ethos for the whole area of international governance of new science and technology whereby the market would play a much more balanced role in determining our political and ethical orientation towards health care.[86] The regime would be driven by human needs. As the former UN Secretary-General Kofi Annan put it, 'The pursuit of development, the engagement with globalisation, and the management of change must all yield to human rights imperatives rather than the reverse'.[87] Amartya Sen distinguishes the hard and soft approaches to development, showing how both are interrelated and how the arguments to espouse the hard approach first are detrimental to peoples' development. In the following passage, he characterises both approaches:

> One view sees development as a 'fierce' process, with much 'blood, sweat and tears' – a world in which wisdom demands toughness. In particular, it demands calculated neglect of various concerns that are seen as 'soft-headed'.... the temptations to be raised can include having social safety nets that protect the very poor, providing social services for the population at large, departing from rugged institutional guidelines in response to identified hardship, and favouring – 'much too early' – political and civil rights and the 'luxury' of democracy. These things, it is argued in this austere attitudinal mode, could

> be supported later on, when the development process has borne enough fruit; what is needed here and now is 'toughness and discipline'. The different theories [diverge] in pointing to distinct areas of softness that are particularly to be avoided, carrying from financial softness to political relaxation, from plentiful social expenditures to complaisant relief. This hard knocks attitude contrasts with an alternative outlook that sees development as essentially a 'friendly' process. Depending on the particular version of this attitude, the congeniality of the process is seen as exemplified by such things as mutually beneficial exchanges (of which Adam Smith spoke eloquently), or by the working of social safety nets, or of political liberties, or of social development – or some combination or other of these supportive activities.[88]

The RTD, within a CHM framework, would reconcile economic interests and human rights, and might produce a pragmatic approach to health.

5.8 Conclusion

This chapter has sought to provide a moral justification for the notion of the CHM and the RTD. In the process, it has also delineated a moral definition of the RTD. The intention was to show that there is a moral reason why the RTD is important and has to be a part of the debate on governance of the human genome. The CHM notion has to be moulded to apply to the human genome, which is a unique resource. There are compelling moral reasons why the different elements of the CHM regime are actually moral elements and therefore why they should be included in any governance debate on the human genome. Various Gewirthian concepts such as reasonable self, ethical individualism and productivist welfarism have been presented here to show the depth of the Gewirthian and the applications that they are susceptible to. The aim of the chapter was to present a case to the reader that the RTD and the CHM regime, as they are found in other governance regimes, are valid and moral, and that they therefore should be treated as viable governance mechanism, based on human rights, to be discussed and adopted in the context of international governance of the human genome.

Notes

1 D. Warner, 'An Ethics Of Human Rights: Two Interrelated Misunderstandings', *Denver Journal of International Law and Policy* 24, 1996, 395–415 at 413.
2 For an elaboration and defence of the law's 'intrinsic moral features', see J. B. Elshtain, 'Law and Moral Life', *Yale Journal of Law and Humanities* 11, 1999, 383–413.
3 H. Charlesworth, *Writings in Rights: Australia and the Protection of Human Rights*, Sydney: University of New South Wales Press Limited, 2002, p. 43.
4 R. Wilde, 'Quis Custodiet Ipsos Custodes?: Why and How UNHCR Governance of "Development" Refugee Camps Should be Subject to International Human Rights Law', *Yale Human Rights and Development Law Journal* 1, 1998, 107–28 at 114; E. A. Engle, 'The Transformation Of The International Legal System: The Post- Westphalian Legal Order', *Queensland Law Review* 23, 2004, 23–45. See also N. Subramanian, L. Chung, 'Measuring The Evolvability of Legal Personality', *IUS Gentium* 11, 2005, 79–130 and

J. D. Ohlin, 'Is the Concept of the Person Necessary for Human Rights?', *Columbia Law Review*, 105, January 2005, 209–49.

5 E. Schrödinger, *What is Life?: The Physical Aspect of the Living Cell*, Cambridge: Cambridge University Press, 1944.

6 A. A. C. Trindade, 'The Contribution of International Human Rights Law to Environmental Protection, with Special Reference to Global Environmental Change' in E. B. Weiss (ed.), *Environmental Change and International Law: New Challenges and Dimensions*, The UN University, Tokyo: UN University Press, 1992 (eBook available online at http://www.unu.edu/unupress/unupbooks/uu25ee/uu25ee00.htm).

7 Despite the numerous advantages of Gewirthian theory it also has its disadvantages. If, as I claim, the appeal of Gewirthian theory is fundamental, the value placed on action could inhibit any positive effects that could be derived from placing value on non-action. Cultures which value non-action as a pathway to truth could be affected by a Gewirthian theory-based legal regime. Are there enough safeguards within the theory to accommodate notions of non-action? Taoism, Buddhism and Hinduism are religions where non action is central to human endeavour. See *Religion and Ethics – Taoism*, available online at http://www.bbc.co.uk/religion/religions/taoism/taoethics/ethics_1.shtml (accessed 29 September 2006); *The Mystical Non-Action*, available at http://jaan.kaplinski.com/philosophy/action.html (accessed 29 September 2006).

8 See J. J. Shestack, 'The Philosophic Foundations of Human Rights', *Human Rights Quarterly* 20 (2), 1998, 201–34 at 225. In his article, Shestack mentions that Alan Gewirth's work merits study. See also G. J. Walters, *Human Rights in an Information Age: A Philosophical Analysis*, Toronto: University of Toronto Press, 2001, p. 36. Walters uses a Gewirthian framework of rights to analyse issues in cyber space and the digital information highway.

9 S. Ridder, L. Woll, 'Transforming The Grounds: Autonomy And Reproductive Freedom', *Yale Journal of Law and Feminism* 2, 1989, 75–98; M. Sunder, 'Piercing the Veil', *Yale Law Journal* 112, 2003, 1399–1472 at 1401.

10 See T. Meron, 'On A Hierarchy Of International Human Rights', *American Journal of International Law* 80, January 1986, 1–23; K. N. Wuerffel, 'Discrimination Among Rights?: A Nation's Legislating a Hierarchy of Human Rights in the Context Of International Human Rights Customary Law', *Valparaiso University Law Review* 1998, 369–412; D. Otto, 'Rethinking The "Universality" Of Human Rights Law', *Columbia Human Rights Law Review* 29, 1997, 1–46; G. M. Wilner, 'Reflections on Regional Human Rights Law', *Georgia Journal of International and Comparative Law*, 25, 1995/1996, 407–26; R. Rajagopal, 'The Allure Of Normativity', *Harvard Human Rights Journal* 11, 1998, 363–77; and U. Baxi, 'Voices of Suffering and the Future Of Human Rights', *Transnational Law and Contemporary Problems* 8, 1998, 125–69.

11 T. A. Aguda, 'Human Rights and the Right to Development in Africa', *The Nigerian Institute of International Affairs*, Lecture Series No. 55, 1989, p. 7.

12 O. Schachter, 'Human Dignity as a Normative Concept', *The American Journal of International Law* 77, 1983, 848–54 at 853.

13 D. A. Hyman, 'Does Technology Spell Trouble with a Capital "T"?: Human Dignity and Public Policy', *Harvard Journal of Law and Public Policy* 27, 2003, 3–18.

14 R. Dworkin, *Taking Rights Seriously*, Boston: Harvard University Press, 1977, p. 198.

15 D. G. Reaume, 'Discrimination And Dignity', *Louisiana Law Review* 63, 2003, 645–95 at 646.

16 G. J. Walters, *Human Rights in an Information Age: A Philosophical Analysis*, Toronto: University of Toronto Press, 2001, p. 35.

17 H. Charlesworth, *Writings in Rights: Australia and the Protection of Human Rights*, Sydney: University of New South Wales Press Limited, 2002, p. 46.

18 A. Gewirth, *Reason and Morality*, Chicago: University of Chicago Press, 1978, p. 290.

19 See The UN Office of the High Representative for the Least Developed Countries, Landlocked Developing Countries and the Small Island Developing States (UN-OHRLLS), where countries are treated as 'development units' and the office of the High Representative seeks to 'assist the Secretary-General in ensuring the full mobilization and coordination of all parts of the United Nations system, with a view to facilitating the coordinated implementation of and coherence in the follow-up and monitoring of the Programme of Action for Least Developed Countries at country, regional and global levels'. Available online at http://www.un.org/special-rep/ohrlls/ohrlls/aboutus.htm (accessed 20 September 2006).

20 R. A. Bohrer, 'Foreword: What is Biotechnology and Why Devote a Law Review Symposium to Biotechnology Law?', *University of Pittsburgh Law Review* 55, 1994, 607–10; F. A. Gevurtz, 'Biotechnology and the Law', *McGeorge Law Review* 32, 2000, 85–7; Senator A. Gore, 'Planning a New Biotechnology Policy', *Harvard Journal of Law and Technology* 5, 1991, 19–30.

21 A. Gewirth, *The Community of Rights*, Chicago: University of Chicago Press, 1996, p. 1.

22 Ibid, p. xiv. See also A. Sen, *Development as Freedom*, Oxford: Oxford University Press, 1999.

23 R. Copelon, 'The Indivisible Framework of International Human Rights: A Source of Social Justice in the U.S.', *New York City Law Review*, May 3, 1998, 59–80; J. E. Bond, 'International Intersectionality: A Theoretical and Pragmatic Exploration of Women's International Human Rights Violations', *Emory Law Journal* 52, 2003, 71–152.

24 *Case Concerning the Barcelona Traction, Light and Power Company Limited, Belgium v. Spain*, 1970, International Court of Justice, p. 3.

25 For a more recent critique see R. Alexy, 'Effects of Defects – Action or Argument? Thoughts about Deryck Beyleveld and Roger Brownsword's Law as a Moral Judgment', *Ratio Juris* 19 (2), 2006, 169–79.

26 See 'Edward Regis, Deryck Beyleveld, Klaus Steigleder' in A. Gewirth, *The Community of Rights*, Chicago: University of Chicago Press, 1996, p. xi.

27 J. Bhabha, 'The Right To Community?', *University of Chicago Law Review* 64, 1997, 1117–28.

28 D. Beyleveld, *The Dialectical Necessity of Morality, An Analysis and Defense of Alan Gewirth's Argument to the Principle of Generic Consistency*, Chicago: Chicago University Press, 1991, p. 184.

29 T. P. Terrell, 'Flatlaw: An Essay on the Dimensions of Legal Reasoning and the Development of Fundamental Normative Principles', *California Law Review* 72, 1984, 288–342 at 329.

30 A. Bertelli, 'Impoverished Liberalism: Does the New York Workfare Program Violate Human Rights?', *Buffalo Human Rights Law Review* 5, 1999, 175–209 at 192.

31 D. Beyleveld, *The Dialectical Necessity of Morality, An Analysis and Defense of Alan Gewirth's Argument to the Principle of Generic Consistency*, Chicago: Chicago University Press, 1991, p. 184.

32 Anonymous, 'Human Rights in the Horn of Africa', *Indiana International and Comparative Law Review* 11, 2001, 543–61 at 544.

33 P. B. Edelman, 'The Next Century of our Constitution: Rethinking Our Duty to the Poor', *Hastings Law Journal* 39, 1987, 1–61 at 22.

34 S. L. Kahn, 'The Right to Adequate Treatment Versus the Right to Refuse Antipsychotic Drug Treatment: A Solution To The Dilemma Of The Involuntarily Committed Psychiatric Patient', *Emory Law Journal* 33, 1984, 441–85 at 475–6; S. R. Salbu, 'Aids and Drug Pricing: In Search of a Policy', *Washington University Law Quarterly* 71, 1993, 691–734 at 712.

35 A. Gewirth, *Human Rights: Essays on Justification and Application*, Chicago: University of Chicago Press, 1982, p. 56.

36 D. Beyleveld, R. Brownsword, *Human Dignity in Bioethics and Biolaw*, Oxford: Oxford University Press, 2001.

37 I. Kant, 'The Doctrine of Virtue', in *The Metaphysics of Morals*, 1797, M. J. Gregor (trans.), 1964, p. 117, paragraph 30, cited in H. J. Steiner and P. Alston, *International Human Rights in Context: Law, Politics and Morals*, Texts and Materials, 2nd edn, Oxford: Oxford University Press, 2000, 262–3.
38 See A. Fagan, *Human Rights*, *The Internet Encyclopaedia of Philosophy*, J. Fieser, B. Dowden (eds), available online at http://www.iep.utm.edu/ (accessed 29 September 2006).
39 W. Hohfeld, 'Some Fundamental Legal Conceptions as Applied in Judicial Reasoning' in W. Cook (ed.), *Fundamental Legal Conceptions as Applied in Judicial Reasoning: Essays by Wesley Newcomb Hohfeld*, New Haven: Yale University Press, 1923, cited in 'Rights', *The Stanford Encyclopedia of Philosophy*, available online at http://plato.stanford.edu/entries/rights/#2.1 (accessed 29 September 2006).
40 A. Gewirth, *Reason and Morality*, Chicago: University of Chicago Press, 1978, p. 1.
41 J. W. Singer, 'The Legal Rights Debate in Analytical Jurisprudence from Bentham to Hohfeld', *Wisconsin Law Review* 1982, 975–1059 at 995–7.
42 This is a restatement of the Principle from Beyleveld and Brownsword, *Human Dignity in Bioethics and Biolaw*, Oxford: Oxford University Press, 2001, pp. 73–4.
43 A. Gewirth, *Reason and Morality*, Chicago: University of Chicago Press, 1978, p. 19.
44 R. Howard, 'Dignity, Community and Human Rights', in A. An-Na'im (ed.) *Human Rights in a Cross-Cultural Perspective*, 1991, p. 81. Sections of this essay are reproduced in H. J. Steiner, P. Alston, *International Human Rights in Context: Law, Politics and Morals: Texts and Materials*, 2nd edn, Oxford: Oxford University Press, 2000, p. 399.
45 H. Lauterpacht, *International Law and Human Rights*, 1950, p. 61, cited in H. J. Steiner, P. Alston, *International Human Rights in Context: Law, Politics and Morals: Texts and Materials*, 2nd edn, Oxford: Oxford University Press, 2000, p. 149.
46 A. Gewirth, 'The Basis and Content of Human Rights', *Georgia Law Review* 13, 1979, 11–43.
47 D. Beyleveld, R. Brownsword, *Human Dignity in Bioethics and Biolaw*, Oxford: Oxford University Press, 2001, p. 82.
48 Ibid., p. 81.
49 N. Lenoir, 'Universal Declaration on the Human Genome and Human Rights: The First Legal and Ethical Framework at the Global Level', *Columbia Human Rights Law Review* 30, 1999, 537–87; see also S. Pridan-Frank, 'Human-Genomics: A Challenge to the Rules of the Game of International Law', *Columbia Journal of Transnational Law* 40, 2002, 619–76 at 641.
50 A. Gewirth, *Reason and Morality*, Chicago: University of Chicago Press, 1978, p. 40.
51 M. Cranston, 'Are There Any Human Rights?', *Daedalus* 112, 1983, 12–13, cited in H. J. Steiner, P. Alston, *International Human Rights in Context: Law, Politics and Morals: Texts and Materials*, 2nd edition, Oxford: Oxford University Press, 2000, p. 185.
52 A. McFarlane, 'In the Business of Development: Development Policy in the First Two Years of the Bush Administration', *Berkeley Journal of International Law* 21, 2003, 521–47 at 522.
53 A. Gewirth, *Reason and Morality*, Chicago: University of Chicago Press, 1978.
54 Ibid., p. 83.
55 Ibid.
56 Ibid., p. 93.
57 Ibid., p. 117.
58 Ibid., p. 109.
59 Ibid., p. 71.
60 Ibid., p. 133.

61 Ibid., p. 135.
62 Ibid., p. 166.
63 D. Beyleveld, R. Brownsword, *Human Dignity in Bioethics and Biolaw*, Oxford: Oxford University Press, 2001, p. 185.
64 Ibid.
65 D. Price, 'The Human Tissue Act 2004', *The Modern Law Review* 68 (5), 2005, 98.
66 D. Beyleveld, R. Brownsword, *Human Dignity in Bioethics and Biolaw*, Oxford: Oxford University Press, 2001, p. 178.
67 A. K. Haas, 'The Wellcome Trust's Disclosures of Gene Sequence Data into the Public Domain and the Potential for Proprietary Rights in the Human Genome', *Berkeley Technology Law Journal* 16, 2001, 145–64 at 151.
68 E. Marshall, 'Bermuda Rules: Community Spirit, with Teeth', *Science* 291 (5507) February 2001, 1192, available online at http://www.sciencemag.org/cgi/content/full/291/5507/1192 (accessed 12 September 2006).
69 A. K. Haas, 'The Wellcome Trust's Disclosures of Gene Sequence Data into the Public Domain and the Potential for Proprietary Rights in the Human Genome', *Berkeley Technology Law Journal* 16, 2001, 145–64 at 164.
70 The idea is to curtail their freedom of action by necessitating that they cooperate, so that other States and individuals of these States can exercise their freedom to participate. For an elaboration of a similar idea, but in the context of a world government, see T. Brühl and V. Rittberger, 'From international to global governance: Actors, collective decision-making, and the United Nations in the world of the twenty-first century' in V. Rittberger (ed.), *Global Governance and the United Nations System*, Tokyo: UN University Press, 2002, p. 25.
71 A. Gewirth, *The Community of Rights*, Chicago: The University of Chicago Press, 1996, p. 109.
72 Article 12(b), Universal Declaration on the Human Genome and Human Rights, UNESCO Doc. C/RES/29/16, 1997.
73 Ibid., Section E.
74 M. G. Erasmus, *The New International Economic Order and International Organisations: Towards a Special Status for Developing Countries?*, Frankfurt/Main: Haag und Herchen, 1979, p. 90.
75 A. Sen, *The Right Not to be Hungry* in P. Alston, K. Tomasevski (eds) *The Right to Food*, Netherlands: SIM, 1984.
76 N. Daniels, 'Technology and Resource Allocation: Old Problems in New Clothes', *Southern California Law Review* 65, 1991, 225–40, at 228.
77 H. Shue, *Basic Rights: Subsistence, Affluence, and U.S. Foreign Policy*, Princeton: Princeton University Press, 2nd edn, 1996.
78 A. Gewirth, *Reason and Morality*, Chicago: University of Chicago Press, 1978, p. 132.
79 Ibid., p. 133.
80 See The Right to Development, Commission on Human Rights RES/59/79, 1999, adopted without a vote.
81 A. Gewirth, *Reason and Morality*, Chicago: University of Chicago Press, 1978, p. 132.
82 Ibid., p. 109.
83 See J. P. Ruger, 'Toward a Theory of a Right to Health: Capability and Incompletely Theorized Agreements', *Yale Journal of Law and the Humanities*, 18, 2006, 273–326, in which she argues that the right to health should be seen as the 'as an ethical demand for equity in health and the need for the internalization of public moral norms to progressively realize this right' (326).
84 A. Gewirth, *Reason and Morality*, Chicago: University of Chicago Press, 1978, p. 109.
85 For example, with respect to debt, the former UN High Commissioner for Human Rights Mary Robinson noted that resources freed from debt payment could go towards

the restoration of health and education systems, which are in a dire state. M. Robinson, 'Constructing an International Financial, Trade and Development Architecture: The Human Rights Dimension' in H. J. Steiner, P. Alston, *International Human Rights in Context: Law, Politics and Morals: Texts and Materials*, 2nd edn, Oxford: Oxford University Press, 2000, p. 1313.

86 N. Daniels, Justice, Healthcare and the Trend Towards Predictive Medicine, *Brussels Symposium : Brocher Foundation*, 22 & 23 November 2004 – Brussels. Available online at: http://www.brocher.ch/archives1/symposium1_en.htm (Accessed 11th February 2009)

87 UN Doc. A/RES/54/1, 1999, at paragraph 275.

88 A. Sen, *Development as Freedom*, Oxford: Oxford University Press, 1999, p. 35.

6 The common heritage of mankind (CHM) and intellectual property rights (IPR)

6.1 Introduction

The twenty-first century has been dubbed the 'century of biology',[1] with biology leading the way to revolutionary changes in society. Backing such change would be the innovation system that fuels the growth of technology – IPR. This is one of the main premises for IPR systems. But IPR protection and public policy goals are both important. Public trust and support are essential for the funding and application of science and technology.[2] Therefore both equitable growth and a rights protection system are essential to create a revolution in biology. This chapter deals with some of the tensions and problems that have arisen in the debates on governance of human genome research in the context of the CHM and IPR. The objective of this chapter is to examine links between international law, human rights, health, development and notions of property, including intellectual property (IP), and to reorient the debate on the CHM and IP.

6.2 International law and intellectual property (IP) law

IP laws are predominantly national in nature and are enforced within national boundaries, through municipal courts.[3] But as IP protection has become more important in international trade there is now more demand for protection of IPR at an international level, across borders.[4] The establishment of the Agreement on Trade-Related Aspects of Intellectual Property Rights (TRIPS) in 1994 institutionalised this link. International trade rules are now issues of importance to nation-states since economic progress and international trade are playing an ever greater role in the progress of a nation. As international trade strengthens and spreads, the nature of interaction between States is also moving away from merely the traditional ways of contact, i.e. through war, border disputes, cultural contacts, etc., to being increasingly occupied by '*lex mercatoria*' issues.

The nature of public international law is consequently changing. It is within this context that international bioethical law will evolve and develop. It is also within this context that we should examine the concept of the CHM and IPR. It used to make sense to understand the notion of the development of the 'law among nations', in the context of the release of European States from the authority of the

Pope through the Treaty of Westphalia, which the Papacy considered to be 'a public act of disregard of the international authority of the Papacy'.[5] Globalisation of trade and economic progress has necessitated the involvement of States in issues other than the quest for independence from the Church and the assertion of sovereign equality between nation-states.[6] However, the issue of the Church and involvement of religion is one that is returning to importance in the context of the governance of genomics.[7]

The law among nations as a law operating *between* rather than *above* states prevailed after the Treaty of Westphalia. In today's world, there is no bigger sovereign than the nation-state, although there are of course powerful states and less powerful states. The purpose of State interaction is to more directly affect the growth and prosperity of peoples by securing cross-border concessions and cooperation to achieve this. In the writings of scholars such as Vittoria, Saurez and Gentili, one can discern the struggle to find the basis for international law after states gained their freedom from the authority of the Church.

We have entered a new phase of international law, where the welfare of the people and communities plays a crucial role in the formation of rules that constitute public international law.[8] This is a phase that is characterised by cooperation rather than just coexistence.[9] The growth of human rights law, international refugee law, environmental law, humanitarian law and trade law is evidence of this trend. The notion of the CHM seeks to take this a step further by synthesising interests of peoples around the world under the banner of humanity.[10] Just as powerful State interests impede the interests of non-powerful States, under modern conditions of private-public partnership in the formation of rules of public international law there are certain private interests that play a similar role to powerful States. Often they derive their power and share it with the powerful States. It is in this context that IPR as a system of monopolies with the potential of abuse by powerful private and State interests that the debate on natural resource exploitation has to be considered in the twenty-first century.

Until recently IPR issues were rarely discussed by the international community,[11] but ever since trade and IP have been linked the international community has been continuously engaged in IPR debates and regulation.[12] The international framework moves much further than the national laws on patents and takes into consideration issues that relate to the welfare of the peoples in its various aspects. In this sense it is converging with the objectives of modern public international law and its expanding sub-disciplines.

6.3 The importance of IPR

From tentative beginnings, when law was first framed to protect the author's rights, IPR law has now grown to include patents, copyright, trademarks, trade secrets and other allied rights. Patents play an especially important role in industrial growth and development, including in the pharmaceutical and biotechnology sector.[13] The move for more protection and a wider coverage of subject matter is strengthening all the time.

IP initially began as a driver of science and technology; it now plays a central role in trade and commerce as the process of globalisation has ushered in the knowledge economy, where innovation and technology play a significant role.[14] It is especially important in biotechnology because of the stage of development that that field is at. Biotech patents form the main assets of companies and are at the heart of the new genetics.[15] Biotechnology and genomics are technology-driven and patent-oriented. Research in the human genome and its applications are being carried out in an environment where public and private spending is equally important to produce health benefits from such research.

There is both a moral and economic argument for IPR. Morally it is right to provide control to individuals over their own creations. Patents, trademarks, copyright, design, plant variety protection, etc. are various types of protection granted to IP. The two sides of the coin in IP protection include protection of producers' interests and the consumers' interests. On the one hand monopoly rights are granted to inventors who create new ideas and objects; on the other hand society derives benefit from such creations. The requirement of disclosure is important – otherwise society cannot derive benefit from such a creation. However does the society's benefit lie narrowly in disclosures or should the creations be beneficial to that society and in that time, by being responsive to the needs of the society? As IPR are harmonised across the world these are issues that are coming into focus. As countries and governments put resources, including legislative, administrative and executive resources into action to strengthen their IP systems, the promise that such systems might hold to respond to the challenges in their respective societies would be worth finding out/debating. Are monopolistic rights suitable within cultures where knowledge production and holding is more non-atomistic, and would innovation thrive better in the absence of such a system of rights protection? The underpinning objectives of the internationalisation of IP protection have to be examined in order to ensure a dialogue between public and private interests and to bring about a balance between the both.

Author rights, which are recognised in various international instruments, stem from the idea that authors have a right to the work they create, and therefore others should not be able to gain monetary advantage from it by copying and reproducing their works. The 1886 Berne Convention for the Protection of Literary and Artistic Works[16] was followed by the 1952 Universal Copyright Convention[17] and the 1961 Rome Convention which seek to protect authors' rights within the territories of the signatory States. The International Covenant on Economic, Social and Cultural Rights recognises authors' rights in Article 15.1. It recognises the right of everyone 'to benefit from the protection of the moral and material interests resulting from any scientific, literary or artistic production of which he is the author'. Protection of industrial property, such as designs and technological innovations, was first provided at the international level through the Paris Convention for the Protection of Industrial Property in 1883.[18]

The economic argument for IPR is clearly applicable to industrial property. Patents, trademarks and other kinds of protection such as industrial designs and trade secrets are types of IPR that protect industrial property. Patents are especially important, and are granted under the following criteria:[19]

The claimed invention must be eligible for patenting.

1 It must be novel.
2 It must be inventive[20] or non-obvious.[21]
3 It must be useful or have industrial application.
4 It must be fully disclosed in the patent application.

Each of these criteria can be interpreted narrowly or broadly depending on the interpretation of the aims and objectives of the system in that particular jurisdiction. The international system of patenting that seeks to harmonise patent laws has come into existence with the coming into force of, TRIPS. Art 27 of TRIPS explicitly lists the criteria for patenting, mentioning utility, inventiveness and novelty and indirectly refering to the eligibility criteria. It allows for patenting of a wide range of inventions, in all fields and states, in Article 27(2):

> Members may exclude from patentability inventions, the prevention within their territory of the commercial exploitation of which is necessary to protect *ordre public* or morality, including to protect human, animal or plant life or health or to avoid serious prejudice to the environment, provided that such exclusion is not made merely because the exploitation is prohibited by their law.

It is advised that states should reconsider their already existing laws if they come into conflict with this provision. *Ordre public* concerns the fundamentals from which one cannot derogate without endangering the institutions of a given society.[22] According to this provision, TRIPS further states that Members may also exclude from patentability:

a diagnostic, therapeutic and surgical methods for the treatment of humans or animals;
b plants and animals other than micro-organisms, and essentially biological processes for the production of plants or animals other than non-biological and microbiological processes. However, members shall provide for the protection of plant varieties either by patents or by an effective *sui generis* system or by any combination thereof.

6.3.1 Problems with IPR

The public policy rationale behind IPR is that encouragement of innovation and creativity will bring benefit to society. The globalisation of IP is beset with problems.[23] For example, although the IP system is said to have encouraged the flourishing of the digital age and is revolutionising music and other cultural sectors, it is also seen as being responsible, in part, for the creation of a 'digital divide'.[24] It seems that both the greater good of the society and the progress of science and technology are being hindered by IP laws. In the Napster case[25], nearly 75 million users everyday were disadvantaged when the website was shut down.[26] Napster enabled users to download MP3 file-format music in permutations and combinations that were not available elsewhere. It was possible for people to download freely single songs without having to buy a whole album. Rather than voicing widespread approval of the action against Napster, commentators questioned the application of the law. One commentator raises the question 'at what point should the law adapt to economic and cultural reality?'[27] The other problem, which is that of stifling innovation, is exemplified in the case of biotechnology patents where initial investments in research are being recouped from patent protection and royalties charged from their use. The case of the breast-cancer gene patent is illustrative of the problem.[28] Patents granted over diagnostic methods involving the breast cancer-genes (BRCA1 and BRCA2) by the US Patent Office created a monopoly over testing for susceptibility for cancer, and it prompted the European Parliament to issue a resolution[29] expressing dissatisfaction with such patents. The resolution implies that such patents over gene sequences would be considered against *ordre public* and morality. The opposition filed at the European Patent Office brings out the various ethical and legal issues involved in granting such patents. The concern is whether this would make research 'so expensive as to eviscerate the very purpose of intellectual property law all together, which is to encourage innovation?'.[30] It has been claimed that issuing patents on gene sequences [in the US] 'is problematic and is harming innovation and denying health care to its citizens as a result of limited access to research and medical diagnostics.'[31] The same can be said of such practices elsewhere in the world. IPR, in many instances, are not optimally serving the needs of the society as a whole.

Individual rights are increasingly at risk because of genomics research.[32] Concerns arising from the lack of protection of traditional knowledge in modern IP law are not uncommon. Human rights concerns have been introduced into the IP law through the exclusion from patentability, based on public health concerns. Natural law theorists of international law would be able to justify the exclusion from patentability those inventions that go against morality.[33] The mandate of the Inter-Agency Committee on Bioethics is to look into cross-cutting issues such as IP related to biotechnology. The tension between IPR and genome research can be seen in the provisions of the UNESCO Declarations and in European legislation. Because of ethical, moral and legal reasons, there are constraints on obtaining IPR on the human genome. According to Article 4 of the 1997 Universal Declaration on the Human Genome and Human Rights, 'the human genome in its natural state shall not give rise to financial gains'. The Council of Europe's 1999 European Convention on Human Rights and Biomedicine similarly declares that 'the human body and its parts shall not, as such, give rise to financial gains'.[34] The Bermuda Rules, which were formed by policy makers and scientists involved in genomic research, involve data-sharing policies for community use, whereby access to information relating to gene sequences can be quickly and freely obtained by interested users all around the world.[35] These guidelines have been formulated to stay the operation of patent law on gene sequences. Despite these rules, however, many patents have been granted on partial DNA sequences,[36] and it is immensely controversial whether such partial DNA sequences satisfy the criteria set out for patentability. The controversy is over whether a mutation of a gene that is already in the public domain should be patentable.[37] The growing number of patents granted to gene sequences highlight the importance of IPR in gene sequences to trade and the economy. It is also apparent that this trend is going to continue.

Fundamental problems with IPR were identified at the Doha Round of the World Trade Organization (WTO) talks, with respect to the protection of public health. Access to life-saving drugs in developing and the least-developed countries is hindered because of patent protection granted to pharmaceutical products.[38] South Africa is one of the countries hard hit by the HIV-AIDS crisis and its government took the drastic step of using the law to bring about compulsory licensing and other patent-waiving measures in order to provide access to life-saving anti-retroviral drugs to its people. The move was opposed by the pharmaceutical association in South Africa and the situation came to symbolise the problems embodied in the TRIPS Agreement and the project to globalise IPR. The South African situation led to the Doha Round of talks in 2001 that resulted in the Doha Declaration on Public Health and TRIPS, which identified the problems of 'commercial and public sector neglect of the R&D needs of developing countries'[39] and the high cost of access to medicines. Use of mechanisms such as compulsory licensing of the production of essential drugs can be made under the TRIPS regime. However, such mechanisms are not suitable to countries which are unable to manufacture such medicines under license because of they lack the industrial capability to do so. TRIPS has further been amended to accommodate the needs of such countries, so that TRIPS Member States can manufacture such drugs and export them to the needy countries. These changes brought about the first proposed amendment of TRIPS. Such a situation highlights the gaps in capabilities of different countries and that different countries require different systems to stimulate innovation. Recognition of this difference is further demonstrated in the staggered entry into force of TRIPS provisions for different countries.

In a report presented to the International Bioethics Committee (IBC) in 2002[40] a list of benefits and concerns that arose from the grant of IPR was discussed. Some fundamental issues were raised, and among the long list of 13 concerns were:

- the change in the tradition of open science
- the change in the balance of public and private research investment
- the character of the genome as being intimate to the human species
- the diversion of research priorities, premature protection
- the novelty requirement
- uncertainty about the genome in its 'natural state'
- 'downstream' use of scientific knowledge for new utility subsequently revealed
- the duration of present IP
- implications for developed and developing countries alike
- equitable benefit sharing and conflicting international rights.

Each of these concerns exemplifies the need for a radical rethink of the rationale and purpose of IP regimes and the role they play in globalised economies.

6.4 IPR and the CHM: a case of neglect

The status of the human genome as the CHM has been considered on many occasions.[41] IPR have come to occupy centre stage in the context of the CHM and its application to the human genome. It is understood, or rather misunderstood, that patenting as a mechanism would be unable to be applied if the human genome were to acquire the legal status of the CHM.[42] The apprehension that such a system would stifle innovation has resulted in the sidelining of the CHM concept in the governance debate. The CHM and IPR have been closely linked since the first meetings of the IBC in 1993. When the IBC began to consider the notion of the CHM as a governing principle in the regulation of the human genome, the concept was introduced with sufficient background on its provenance and its operation towards different natural resources and common spaces. But, following this, the attention shifted from understanding the regime and the operation of it as a whole to those components of it which dealt with ownership. The position in law about ownership and property rights in body parts has been well established – the law does not allow property rights in body parts or samples. However, the position does not seem at all clear in the context of providing patent protection to products using DNA. It seems that it is possible to gain IPR over genes and DNA while a direct property right is prohibited. Partial gene sequences, which can be considered as the natural state of the DNA, and which arguably do not satisfy the conditions of invention within patent law, could be patented.

The normative content of the CHM was not well understood, despite attempts made by some members of the IBC to clarify the CHM notion.[43] It seems that the contentious nature of the ownership involved in any discussion on the CHM precluded its very consideration as a useful tool for governance. In 1994, when the Legal Commission of the IBC met to discuss the form and content of the international instrument for the protection of the human genome, the CHM status was interpreted merely to mean the shared genetic heritage of humanity and that each human being was a depository of his/her own genome and of the genome of the entire species. The discussions did not reflect the idea of CHM as a normative international legal concept, although international law was invoked as the governing law for the human genome.[44] The potential for consideration of 'development' issues and issues pertaining to equity in health research,[45] and how this could be achieved through a governance framework, could have featured much earlier in the IBC if the normative legal concept of CHM had been adopted instead of a rather obvious restatement of the biological status of the human genome.

During the 5th Meeting of the Legal Commission of the IBC in 1995[46] the CHM debate centred on whether the human genome could be subject to individual or group ownership. The report of the meeting stated 'the declaration should emphasize the fact that before being part of common heritage the human genome was first and foremost the property of the individual.' Concerns about free access to results of genome research were part of the debate. The 6th Meeting saw a more nuanced discussion on the term, where patentability of genes and equitable exploitation of the genome were considered. However, it was felt desirable that the notion of CHM be expanded no more than it was in the draft Declaration, since the question of patentability could not be resolved in the IBC without involving other agencies such as the World Intellectual Property Organization (WIPO) and the WTO.[47] The idea that the system of patenting could pose a threat to realising health goals[48] was not a point of discussion.

In the final form of the 1997 Universal Declaration the section on the CHM included the term 'symbolic' to qualify the notion of the CHM in the Universal Declaration, with the intention of reducing its significance as a guiding principle within the Declaration. The intervention of the intergovernmental experts in July 1997 was instrumental in downgrading the status of the CHM from a cornerstone principle to one of symbolic presence within the Universal Declaration. The IBC could have averted such a result if the debates had focused on the normative framework as a whole as opposed to just ownership aspects. Ultimately, the view of the intergovernmental experts that a CHM framework would provide disincentives to innovation prevailed as the strongest argument.

Two reports submitted by IBC Working Groups in 2001 and 2002 on solidarity and the ethics of IP protection addressed important issues on these themes, but only after the 1997 Universal Declaration had been adopted. The IBC report in 2001 on 'Solidarity and International Co-operation between Developed and Developing Countries concerning the Human Genome'[49] considered the comparative advantages and disadvantages of elements that constitute a CHM regime. Only a passing mention was made of Article 1 of the Universal Declaration. In January 2002, a report was submitted to the IBC by Justice Michael Kirby titled 'Ethics, Intellectual Property and Genomics'[50] and focusing on the 'fundamental issue' of securing benefits of human genome research for the whole of humanity. Published just after the human genome sequence had been finished,[51] it envisaged the idea of benefit sharing as having assumed an element of urgency. It welcomed the appointment of the Inter-Agency Committee on Bioethics to discuss cross-cutting issues such as IPR while concluding that 'in accordance with the provision of Article 27(2) of the Trade related Intellectual Property Rights, the human genome is not patentable on the basis of the public interest considerations set out therein, in particular, public order, morality and the protection of human life and health.' It recommended that UNESCO promote urgently the adoption of an international convention on ethical and other issues relating to both IP and genomics, or on living matter including IP and genomics.

As has been discussed in the earlier chapters of this book, the notion of the CHM is primarily concerned with equitable use and with the fulfilment of the right to development (RTD). It does not oppose the granting of property rights and it provides for alienation of resources based on existing property law. At the same time it does not allow appropriation and effectively nullifies the operation of property law in certain areas. Such a scheme of governance under a CHM regime denotes that property law is not central to the CHM; it is equitable use that is the main concern of a CHM regime. Therefore it is not factual that a CHM regime would negate patent holding on genes. Using the terminology of the Nuffield Council Report, on the whole the consequentialist and Lockean justifications[52] for the existence of a patent system are not challenged at the IBC.

6.5 IP and property: morally equal rights?

The moral justifications for IPR do not occupy the same place as that of property rights, although the movement for ever stronger protection of IPR seems to presume so. The notion of property was relatively 'settled', in terms of land rights and private income. But the increase of biological research and the introduction of 'bio property', together with the importance of IP in human genome research, have all challenged the notion of property in a number of ways. With increased reliance on research into diagnostic methods, drugs and treatment modes to address disease for addressing health problems, the idea of property in the biological has been reopened.[53] The importance of biological material from across different countries and indigenous communities has given rise to new questions about communal ownership and non-ownership and non-property, i.e. whether to leave things out of the property realm.[54]

But the justification of IP on the grounds that it is the protection of property rights cannot be supported on a moral basis. In the South African AIDS drugs case,[55] the applicants relied on the right to property guaranteed under Section 25 of the South African Constitution to protect the patents held by the pharmaceutical company. In the petition submitted by the Treatment Action Group (TAG), the group rightly pointed to the ethical priority that should be given to health over property rights. Quoting from the writings of the one of the members of the Bench, TAG argued that human dignity was a foundational value of the South African Constitution and that it included the provision of housing, health care, food, water and social security.[56]

At a moral level, the right to property and IP cannot be considered to be on the same footing.[57] The right to property occupies a fundamental position in the spectrum of rights, as it is instrumental to the promotion of basic rights such as shelter and protection of the self. To the extent that it is essential for the fulfilment of the basic needs of humans, the right to property is a fundamental right. The right to IPR on the other hand is not essential for the fulfilment of the basic needs in the same way as that of the right to property. Even under a Lockean system of rights, fruits of one's labour ought to be protected by the grant of IPR, in return for which 'the public will be served and progress of science and the useful arts will be promoted.'[58] The argument can be posed that if public interest is not served, then the grant of such rights can be withdrawn on the grounds that such rights are detrimental to public interest. In the Gewirthian sense of property rights basic needs would trump the grant of IPR if the fulfilment of basic rights (the right to health) comes into conflict with the fulfilment of additive or non-subtractive rights (the right to have exclusive patent protection). However, further examination of the notion of IPR and its relation to basic rights is required.

The notion of the CHM will not provide any answers or solutions to the problematic notions of property in the biological, but it will serve to impose an overarching condition that should be fulfilled by any system of property rights, i.e. the promotion of the RTD of all peoples. It might help to bring all these questions together, in order to address them fairly and equitably, with the central focus being the fulfilment of the RTD of all peoples.[59] Human rights are about the universal;[60] the RTD is about enabling individuals acquire the basic potential for development and to be free to make their choices. A system of law laying down rules relating to property should therefore be RTD friendly. It should promote and enable individuals to acquire their basic potential to act – and not act as a hindrance to developing their potential. The CHM concept is also an explicit recognition of the interdependence of States and peoples, which is reinforced in the context of the human genome, which itself has been shaped over the centuries by peoples all around the world.

The call for an integrated intellectual rights system, which WIPO seems to be promoting, comes to the issue of governance from the angle of protection of private property rights.[61] The CHM concept, on the other hand, comes to IPR from the angle of the protection of the RTD. The rationale behind WIPO and its central project has many powerful supporters,[62] unlike the CHM notion, which is relatively weak and at present does not have a central place in the international debates on the governance of the human genome. Support for it is fragmented; mainly coming from civil society[63] and it is not looked upon as a pragmatic framework of governance by the main actors. Grotius used it to great effect; Pardo was less successful in embedding it into the law of the sea. It remains to be seen how the notion of the CHM will withstand the test of time in the twenty-first century – will it be a mere ornament, serving no real purpose, or a robust regime presiding over resource allocation and property issues?

6.6 WIPO policy and the Development Agenda

A fundamental rethink of the agenda of IPR protection is essential in order to address concerns of an equitable nature. Such a reform would be welcome in the wake of burgeoning technological advancements, including human genomic research. The WIPO development initiative is a step in this direction.

In 2004 Brazil and Argentina, supported by developing countries,[64] submitted a proposal to WIPO to initiate a development agenda for the organisation. They were concerned that the harmonisation of international property law across the world was happening without due regard to the difference in development in various countries. The proposal recalled that the most important agenda for the international community was development. In this context, they referred to the Millennium Development Goals[65], the Monterrey Consensus[66], the Sao Paulo Consensus adopted by UNCTAD XI,[67] and the WTO Doha Round of multilateral trade talks. They urged WIPO to assess 'the role of intellectual property and its impact on development' on a case-by-case basis. They wanted to ensure that 'the "development dimension" is unequivocally determined to constitute an essential element of the Organization's work program' by opening up the possibility of amending the 1967 WIPO Convention. Particular concern was raised about the Draft Substantive Patent Law Treaty (SPLT) in the Standing Committee on the Law of Patents (SCP) because of the provisions to raise patent protections and the creation of new obligations which developing countries will find hard to fulfil. Among the measures recommended were the adoption of a high-level declaration on IP and development, thus making it visible and important, and amending the WIPO Convention to explicitly incorporate the development dimension into WIPO's objectives and functions. The change suggested was the following – it relates to paragraph (i) of Article 3 of the WIPO Convention, dealing with the objectives of the convention:

> (i) to promote the protection of intellectual property throughout the world through cooperation among States and, where appropriate, in collaboration with any other international organization, *fully taking into account the development needs of its Member States, particularly developing countries and least-developed countries.*

Discussions on transfer of technology and technical cooperation were among the other issues that the countries saw as being part of the Development Agenda. The procedural aspects of the Development Agenda included the increased participation of civil society in WIPO activities, the setting up of working groups to consider development issues and joint sessions between WIPO and UNCTAD. An expanded version of the proposal was discussed in the 2005 Inter-sessional Intergovernmental Meeting on a Development Agenda for WIPO (IIM).[68,69] A proposal by the US essentially disagreed with the Development Agenda on the basis that WIPO had had 'a robust "development agenda" in all of its work for a long time, delivering high-quality development activities to Member States on a demand-driven basis.' In the spirit of the proposal submitted by other Members, the US put forward a partnership programme for the coordination of IP-related development assistance from the available sources.[70] It opposed changes to the 1967 Convention on the basis of the 1974 agreement between the UN and WIPO (Article 57, Charter of the UN) stating that this agreement only allocates a subsidiary role to WIPO in development. The agreement recognises WIPO as a Specialised Agency of the UN that is responsible for measures to accelerate economic, social and cultural development subject to the competence and responsibilities of the UN and its organs, particularly the UN Conference on Trade and Development, the UN Development Programme and the UN Industrial Development Organisation, as well as of UNESCO and of other agencies within the UN system.[71]

Quoting the WIPO Director-General Kamil Idris, the proposal states that IP is a powerful tool for economic growth[72] and gives examples of its success in the cultural sector (films, music, literature) and in the technological sector (computer software, information and communication technologies). The UK proposal to the IIM gives cautious support to the proposals of Argentina and Brazil, and to that of the US,[73] and cites the study conducted by the Commission on Intellectual Property Rights in 2003 and the Commission for Africa's 2005 report called *Our Common*

Interest. The UK stated it was not hesitant to amend the WIPO Convention if development issues were being neglected but its position was also that WIPO's existing mandate allows for the effective integrating of development objectives into its activities. It noted that 'Technical cooperation must be tailored to countries' needs and should be seen in a broad sense including enhancing capacity to facilitate the development of balanced IP-related policies.'

The WIPO Development Agenda is still in its infancy and the support and opposition it incurs are both well founded. It states that the genesis of the global harmonisation project has to be revisited and, if necessary, adjusted in order to allay the fears of developing countries that such a system is working to their detriment. It could, if taken in the right direction, recognise and support the promotion of the RTD by creating an institutional framework which recognises 'IP as one of many tools for development – not as an end in itself'.[74] Despite objections and disagreements, the existence of the Development Agenda is a formal recognition of the gaps and inconsistencies in IP law and is essential for the much needed change in 'role and attitude'[75] of the organisation. The success of a CHM regime depends on IP reform and no realistic consideration of such a regime is possible if the globalisation of IP law carries on without IP reform. The reform proposal uses familiar language. The proposed change to the WIPO Convention refers to the development needs of developing and the least-developed countries, identical to the 1982 Convention on the Law of the Sea or the 1979 Moon Treaty and, more recently, the 1896 Declaration on the Right to Development and UNESCO's 1997 and 2005 Universal Declarations.

The technology transfer clause[76] is one of the most important in the reform proposals submitted by WIPO Members. It calls for an effective regime of technology transfer and recommends 'the creation of a Standing Committee on Intellectual Property and the Transfer of Technology, for the consideration of measures to ensure an effective transfer of technology to developing countries and LDCs'.[77] The provision on technical assistance is development-oriented and emphasises the importance of fulfilling pressing global needs. The ubiquitous reference to such goals is seen in Section VII of the Proposal which seeks 'not to overburden scarce national resources that may be more productively employed in other areas', and calls for action to ensure that 'the costs of IP protection do not outweigh the benefits thereof'. The proposal also aims 'to ensure that the social costs of IP protection are kept at a minimum' and that they are 'tailored to meet each country's level of development and are fully responsive to the specific needs and problems of individual societies', 'in particular to promote important public policy objectives'.

The Proposal seems to be moving towards a package where public policy objectives and individual country needs, based on their stage of development, will determine the type of IP protection that would be given within that jurisdiction. At the core of the priorities is development. On the whole, these changes, when applied in the context of human genome research, would shape themselves around the 1997 and 2005 UNESCO provisions relating to technology transfer and international cooperation and solidarity.

In 2006 the 2nd Session of the Provisional Committee on Proposals Related to a

WIPO Development Agenda[78] yielded the recommendation that 'a process for the adoption of a high-level declaration on intellectual property and development'[79] be initiated. The Committee emphasised the importance of social and economic development and challenged the present model of IP via the 1967 WIPO Convention which does not prescribe a particular model of IPR to be followed. It states that 'nothing in the 1967 WIPO Convention prevents WIPO from undertaking any initiative to consider various models of innovation other than Intellectual Property'. It goes on to state that the current efforts at harmonisation of IP are not taking into consideration its effects on social and economic development and are therefore undermining the UN-WIPO Agreement. The document is far reaching and seeks to introduce major changes to IP management and use. The Development Agenda is broad and includes all the majors concerns relating to health, agriculture, traditional knowledge, protection of public domain, transfer of technology, training programmes, integration of competition law into IP, and other issues. The 2006 WIPO Session[80] extended the mandate of the Provisional Committee to further discuss proposals and to streamline different proposals.

In 2008 WIPO formed the Committee on Development and Intellectual Property (CDIP), which is preparing a report on the issue. The CDIP aims to put WIPO in touch with the issues of the twenty-first century.

6.7 Conclusion

A CHM regime is fully capable of accommodating a robust IPR system that would be of service to both public and private interests. To dismiss it by stating otherwise is to take a step away from integrating broader international normative concepts into the governance of the human genome. Such initiatives hinder rather than support the strengthening of international law and the basis of rule of law.

Notes

1 J. Hanke, 'Genomics and New Technologies as Catalysts for Change in the Drug Discovery Paradigm', *Journal of Law, Medicine and Ethics* 28, 2000, 15–22 at 17.
2 B. Wynne, 'Public Engagement as a Means of Restoring Public Trust in Science – Hitting the Notes, but Missing the Music?, *Community Genetics* 9, 2006, 211–20.
3 W. Cornish and D. Llewelyn, *Intellectual Property: Patents, Copyright, Trademarks and Allied Rights*, London: Thompson, Sweet and Maxwell, 5th edn, 2003, p. 26.
4 G. B. Dinwoodie, 'The International Intellectual Property Law System: New Actors, New Institutions, New Sources', *Marquette Intellectual Property Law Review Special Issue* 10, 2006, 205.
5 L. Gross, *The Peace of Westphalia, 1648–1948* in *International Law in the Twentieth Century*, cited in L. Gross (ed.), *The American Society of International Law*, New York: Appleton-Century-Crofts, 1969, p. 33.
6 Ibid.
7 M. Arsanjani, 'Negotiating the UN Declaration on Human Cloning', *The American Journal of International Law* 100:1, January 2006, 164–79 at 167.
8 Earlier, even when treaties were made to safeguard the interests of individuals (e.g. treaties against slave trade which were founded to a great extent on humanitarian considerations), they were also motivated by the self-interest of the contracting states.

See A. Cassese, *International Law*, 2nd edn, Oxford: Oxford University Press, 2005, p. 376.

9 B. F. Fitzgerald, 'Trade-based Constitutionalisms: The Framework for Universalizing Substantive International Law?', *University of Miami Yearbook of International Law* 97:5, 1996, 111–63 at 116.

10 C. W. Jenks, 'Craftsmanship in International Law', in *International Law in the Twentieth Century* in L. Gross (ed.), *The American Society of International Law*, New York: Appleton-Century-Crofts, 1969, p. 77, where Jenks talks about the transformation of international law from being the public law of Europe to being the law of a universal world community.

11 F. Abbott, T. Cottier, F. Gurry, *The International Intellectual Property System: Commentary and Materials, Part I*, The Hague: Kluwer Law International, 2003, p. 502.

12 See TRIPS and the Doha Development Agenda (DDA) at *Doha Development Agenda: Negotiations, implementation and development*, World Trade Organisation, http://www.wto.org/english/tratop_e/dda_e/dda_e.htm (Accessed 28th June 2008). Key documents include *Doha Ministerial Declaration*, WT/MIN(01)/DEC/1, 20 November 2001 and Declaration on the TRIPS agreement and public health WT/MIN(01)/DEC/2, 20 November 2001

13 N. N. Aljalian, 'The Role of Patent Scope In Biopharmaceutical Patents', *Boston University Journal of Science and Technology Law* 11, 2005, 1–76 at 1.

14 D. J. Skyrme, 'The Global Knowledge Economy: and its implication for markets', *Insight* 21, 1997, available online http://www.skyrme.com/insights/21gke.htm (accessed 10 November 2006); see also Center for Strategic Economic Studies (CSES), *A Primer on the Knowledge Economy*, CSES: Victoria University, 2000, available online at http://www.cfses.com/documents/knowledgeeconprimer.pdf (accessed 16 April 2003).

15 A. K. Haas, 'The Wellcome Trust's Disclosures of Gene Sequence Data into the Public Domain and the Potential for Proprietary Rights in the Human Genome', *Berkeley Technology Law Journal* 16, 2001, 145–64 at 149.

16 Signed 9 September 1886, revised at Paris on July 24, 1971 and amended in 1979, S. Treaty Doc. No. 99–27, 1986.

17 UN Treaty Series No. 13444 (943), pp 178–325.

18 UN Treaty Series No. 11851 (828), pp. 305–88.

19 *The Ethics of Patenting DNA – A Discussion Paper*, Nuffield Council on Bioethics, London, 2002, p. 11.

20 As per the terminology used in European Patent Convention.

21 As per terminology used in US patent law.

22 D. Gervais, *The TRIPS Agreement: Drafting History and Analysis*, London: Sweet and Maxwell, 2nd edn, 2003, p. 222.

23 The emergence of the WTO-TRIPS framework and increased activity towards the harmonisation of IP rules in the World Intellectual Property Organization (WIPO) is referred to as the globalisation of intellectual property. See M. Chon, 'Intellectual Property and The Development Divide', *Cardozo Law Review* 27, 2006, 2821–912 at 2825; R. Weissman, 'A Long, Strange TRIPs: The Pharmaceutical Industry Drive to Harmonize Global Intellectual Property Rules, and the Remaining WTO Legal Alternatives Available to Third World Countries', *University of Pennsylvania Journal of International Economic Law* 17, 1996, 1069–1125, esp. 1072–7.

24 L. Mtima, 'Introduction to Symposium Intellectual Property and Social Justice', *Howard Law Journal* 48, 2005, 571–7 at 572.

25 *A&M Records, Inc. v. Napster*, 114 F. Supp. 2d 896 (N.D. Cal. 2000).

26 M. Wolff, 'Stop, Thief!', *New York Magazine*, 3 March 2003, pp. 24–6.

27 P. J. Gardner, 'Our Intellectual Property World: Charting a Course from Ancient Native Cultures to the Threshold of the Future on a Roiled Sea Of Macroeconomics, Morality And Ethics', *IDEA: The Journal of Law and Technology* 44, 2004, 237–45 at 239.

28 J. Paradise, 'European Opposition to Exclusive Control Over Predictive Breast Cancer Testing and the Inherent Implications for U.S. Patent Law and Public Policy: A Case Study of the Myriad Genetics' BRCA Patent Controversy', *Food and Drug Law Journal* 59, 2004, 133–154.

29 *European Parliament resolution on the patenting of BRCA1 and BRCA2 ('breast cancer') genes*, Official Journal C 87E, 11.4.2002, p. 265, available online at http://europa.eu.int/eur-lex/pri/en/oj/dat/2002/ce087/ce08720020411en02650266.pdf.

30 Gardner op. cit., p. 240.

31 B. Gargano, 'The Quagmire of DNA Patents: Are DNA Sequences More Than Chemical Compositions of Matter?', *Syracuse Science & Technology Law Reporter* 2005, 3–5.

32 L. B. Andrews, J. Paradise, 'Gene Patents: The Need for Bioethics Scrutiny and Legal Change', *Yale Journal of Health Policy, Law & Ethics* 5, 2005, 403–12 at 412.

33 F. Abbott, T. Cottier, F. Gurry, *The International Intellectual Property System: Commentary and Materials, Part I*, The Hague: Kluwer Law International, 2003, pp. 503–4.

34 Convention for the Protection of Human Rights and Dignity of the Human Being with regard to the Application of Biology and Medicine, CETS No. 168, Council of Europe, available online at http://conventions.coe.int (accessed 29 March 2006).

35 See, for example, the Human Genome Program of the US Department of Energy Office of Science, Summary of Principles Agreed at the First International Strategy Meeting on Human Genome Sequencing, Bermuda, 25–28 February 1996), available online at http://www.ornl.gov/sci/techresources/Human_Genome/research/bermuda.shtml#1; Sharing Data from Large-Scale Biological Research Projects: A System of Tripartite Responsibility, Wellcome Trust, 2003, available online at www.wellcome.ac.uk/assets/wtd003207.pdf; A.K. Haas, 'The Wellcome Trust's Disclosures of Gene Sequence Data into the Public Domain and the Potential for Proprietary Rights in the Human Genome', *Berkeley Technology Law Journal* 16, 2001, 145–64 at 151.

36 *In re Thomas F. Deuel, Yue-Sheng Li, Ned R.Siegel and Peter G. Milner*, No. 94–1202, United States Court of Appeals, Federal Circuit, 28 March 1995, 51 Federal Reporter, 3rd Series, p. 1552.

37 J. Murray, 'Owning Genes: Disputes Involving DNA Sequence Patents', *Chicago-Kent Law Review* 75, 1999, 231–57 at 249; D. Nicol, 'On the Legality of Gene Patents', *Melbourne University Law Review* 29, 2005, 809–45.

38 B. C. Mercurio, 'TRIPS, Patents, and Access to Life-Saving Drugs in the Developing World', *Marquette Intellectual Property Law Review* 8, 2004, 211–53 at 223.

39 E.'t Hoen, 'TRIPS, Pharmaceutical Patents, and Access to Essential Medicines: A Long Way from Seattle to Doha', *Chicago Journal of International Law* 3, 2002, 27–46 at 46.

40 Rapporteur: Justice Michael Kirby, UN Doc. SHS-503/01/CIB-8/2 Rev., 2002.

41 *The Ethics of Patenting DNA*, Nuffield Council of Bioethics, 2002. The Nuffield Council on Bioethics was established in 1991 to consider ethical issues arising from developments in medicine and biology in order to assist policy makers and others. In the case of this report, it was to particularly help the courts, patent lawyers and patent offices. *The Ethics of Patenting DNA* was published in 2002, the members of the Roundtable Meeting which helped prepare this report include Professors Tom Baldwin, Martin Bobrow, John Barton, Sir Brian Heap, Joseph Straus, Dame Merilyn Strathern, Mike Stratton, Alan Williamson and the Hon. Mr Justice Jacob. Also see T. Lezemore, *Should DNA be Patented?*, Presentation during the launch of the Nuffield Council on Bioethics paper *The Ethics of Patenting DNA: A Discussion Paper*, 23 July 2002, The Royal Society, London: Nuffield Council on Bioethics, 2002.

42 L. B. Andrews, J. Paradise, 'Gene Patents: The Need for Bioethics Scrutiny and Legal Change', *Yale Journal of Health Policy, Law & Ethics* 5, 2005, 403–12, where the authors discuss the future of patenting for DNA sequences and express the opinion

that patenting gene sequences is being challenged by the view that genes belong to the CHM.

43 *Proceedings of the Third Session of the IBC*, 1995, available online at http://portal.unesco.org/shs (accessed 15 May 2006), pp. 107–8.
44 Ibid., pp. 47–8.
45 A more general point is made in A. J. London, 'Justice and the Human Development Approach to International Research', *Hastings Center Report* 34 (1), 2005, 24–37.
46 *Report of the Fifth Meeting of the Legal Commission of the IBC*, Paris: UNESCO, 1995, available online at http://www.unesco.org/ibc/en/genome/juridique/r1.html (accessed 24 November 2001), pp. 82–3.
47 *Report of the Seventh Meeting of the Legal Commission of the IBC*, Paris: UNESCO, 1996, available online at http://www.unesco.org/ibc/en/genome/juridique/r1.html (accessed 24 November 2001), pp. 47–8.
48 D. Barnard, 'In the High Court of South Africa, Case No. 4138/98: The Global Politics of Access to Low-Cost AIDS Drugs in Poor Countries', *Kennedy Institute of Ethics Journal* 12 (2), 2002, 159–74.
49 Rapporteur: Mehmet Öztürk, UNESCO Doc. BIO-7/00/GT-2/3 (Rev. 1), 2001.
50 Rapporteur: Justice Michael Kirby, UNESCO Doc. SHS-503/01/CIB-8/2 Rev., 2002.
51 'Editorial: The Human Genome Sequence', *Science* 291 (5507), 2001, 1145–434.
52 The consequentialist view is exemplified by values which, for example, would encourage a system of public rules that provides security and benefits and that can be adapted to serve the public interest (compulsory licensing, etc.). The Lockean view is that patenting can be justified on that basis that one has a 'natural' right to that which one has 'mixed one's labour' with. See S. Thomas, *The Ethics of Patenting DNA*, Nuffield Council on Bioethics, 23 July 2002, Presentation at the launch of the report, slide 11.
53 M. R. Sobota, 'The Price Of Life: $50,000 for an Egg, Why not $1,500 for a Kidney? An Argument to Establish a Market for Organ Procurement Similar to the Current Market for Human Egg Procurement', *Washington University Law Quarterly* 82, 2004, 1225–49.
54 P. Oldham, 'Global Status and Trends in Intellectual Property Claims: Genomics, Proteomics and Biotechnology' submitted as UNEP/CBD/WG-ABS/3/INF/4 by the European Community to the Third Meeting of the Ad-Hoc Open-Ended Working Group on Access and Benefit-Sharing, United Nations Convention on Biological Diversity, Bangkok, Thailand, 14–18 February 2005.
55 *The Pharmaceutical Manufacturer's Association of South Africa and others and the President of the Republic of South Africa, The Honourable Mr N. R. Mandela N. O. and others and Treatment Action Campaign*, Amicus Curiae Case No: 4183/98 (Transvaal Provincial Division), 2001.
56 Justice Chaskalson, 'Human Dignity as a Foundational Value of Our Constitutional Order', *South African Journal on Human Rights* 16, 2000, 201–2.
57 K. W. Baer, 'A Theory of Intellectual Property and the Biodiversity Treaty', *Syracuse Journal of International Law and Commerce* 21, 1995, 259–81 at 260.
58 A. D. Moore, 'A Lockean Theory Of Intellectual Property', *Hamline Law Review* 21, 1997, 65–108 at 66.
59 C. R. Buxton, 'Property in Outer Space: The Common Heritage of Mankind Principle vs. The "First In Time, First In Right" Rule of Property Law', *Journal of Air Law and Commerce* 69, 2004, 692.
60 Here the term universal is used as a principle 'that emphasises the essential worth of every human being whilst allowing for diversity of culture, religion, gender, sexuality and so on' – see L. S. Bell, A. J. Nathan, I. Peleg (eds), *Negotiating Culture and Human Rights*, New York: Columbia University Press, 2001; D. Otto, 'Rethinking the "Universality" of Human Rights Law', *Columbia Human Rights Law Review* 29, 1997, 1–46. See also R. Pannikar, 'Is the Notion of Human Rights a Western

Concept?', *Diogenes* 120, 1982, 75, where he argues that the conception of human rights is a Western conception, but one that the world should not renounce declaring or enforcing.

61 See Intellectual Property Rights and Human rights, Commission on Human Rights RES/07, 1995.

62 S. K. Sell, 'Private Power, Public Law: The Globalisation of Intellectual Property Rights', *Cambridge Studies in International Relations Series*, No. 88, 2003.

63 Treaty to Protect Genetic Commons, available online at http://sbb.collectifs.net/Textes/Rifkin.htm (accessed 13 October 2006).

64 See Proposal by Argentina and Brazil for the Establishment of a Development Agenda for WIPO, WIPO Doc. WO/GA/31/11, 2004, which was c-sponsored by Bolivia, Cuba, Dominican Republic, Ecuador, Egypt, The Islamic Republic of Iran, Kenya, Peru, Sierra Leone, South Africa, United Republic of Tanzania and Venezuela.

65 The MDGs 'form a blueprint agreed to by all the world's countries and all the world's leading development institutions' for furthering the goals of development and are.available online at http://www.un.org/millenniumgoals/ (accessed 12 November 2006).

66 The Monterrey Consensus consists of a collection of non-binding and vague declarations of intent and was formed at the conclusion of the Conference on Financing for Development (FfD) on 22 March 2002. The developing countries commited themselves to good governance, economic and social reforms and stability-oriented macro policies. See ELDIS: The Gateway to Development Information available online at http://www.eldis.org/static/DOC1942.htm (accessed 12 November 2006).

67 In the Sao Paulo Consensus adopted in June 2004, UNCTAD 'reaffirmed its commitment to the eradication of poverty and hunger, and to the achievement of fair and equitable multilateral trade negotiations'. See 'UNCTAD XI Concludes with Adoption of "Sao Paulo Consensus" Focusing on Poverty Reduction in Least Developed Countries', Press Release, 21 June 2004, UNCTAD, available online at http://www.unis.unvienna.org/unis/pressrels/2004/tad1994.html (accessed 12 November 2006).

68 The IIM was constituted to examine the proposals contained in WIPO Doc. WO/GA/31/11 (Proposal by Argentina and Brazil for the Establishment of a Development Agenda for WIPO), as well as additional proposals of Member States.

69 Proposal to establish a Development Agenda for WIPO: An elaboration of issues raised in WIPO Doc. WO/GA/31/11, IIM/1/4, 2005.

70 Proposal by the United States of America for the establishment of a partnership program in WIPO, Inter-sessional Intergovernmental Meeting on a Development Agenda for WIPO First Session, Geneva, 11–13 April 2005, IIM/1/2, 18 March 2005, p. 2.

71 Article 1, Agreement between the United Nations and the World Intellectual Property Organization, 17 December 1974, available online at www.wipo.int/treaties/en/agreement/index.html (accessed 13 March 2005).

72 K. Idris, *Intellectual Property: A Power Tool for Economic Growth*, WIPO, 2003.

73 Inter-sessional Intergovernmental Meeting on a Development Agenda for WIPO, First Session, Geneva, 11–13 April 2005, Proposal by the United Kingdom, IIM/1/5, 2005.

74 See the Summary of the Geneva Declaration on WIPO available online at http://www.cptech.org/ip/wipo/da.html (accessed 12 November 2006).

75 J. Boyle, 'A Manifesto on WIPO and the Future of Intellectual Property', *Duke Law & Technology Review*, 0009, 2004, available online at http://www.law.duke.edu/journals/dltr/articles/2004dltr0009.html (accessed 12 November 2006).

76 Section V and Recommendation 5 deal with transfer of technology – see Proposal by Argentina and Brazil for the Establishment of a Development Agenda for WIPO, WIPO Doc. WO/GA/31/11, 27 August 2004.

77 Ibid., Recommendation 5.

78 WIPO, Second Session, Geneva, 26–30 June 2006

79 Proposal by Argentina, Bolivia, Brazil, Cuba, Dominican Republic, Ecuador, Egypt, the

Islamic Republic of Iran, Kenya, Peru, Sierra Leone, South Africa, Tanzania, Uruguay and Venezuela.

80 'WIPO Member States Agree on a Way Forward for discussions on a WIPO Development Agenda', Press Release 461, Geneva, 2 October 2006, available online at http://www.wipo.int/edocs/prdocs/en/2006/wipo_pr_2006_461.html (accessed 28 October 2006).

7 Conclusion

The international legal governance of the human genome is becoming more fraught with complexity as scientific research pushes ever further ahead. This concluding chapter summarises the arguments made in the previous chapters of this book.

The Human Genome Project (HGP) has the potential to reveal a lot more about ourselves, along with proteomics, stem-cell research, nanotechnology, robotics, etc. and these new endeavours endow us with extraordinary capabilities. Digital and electronic technologies are having a major impact on research and health care. Private and public funding in science and technology has increased expectations of short- and long-term benefits from such activities. The HGP has thrust molecular genetics into the international limelight. Japan was first to start to engage in a sequencing project, but its importance was played down by other powers. Not long after, scientists and technologists in the US were able to convince their health and defence ministries to fund a techno-science project to sequence the human genome. The International Human Genome Organisation was formed and the Bermuda Rules, which called for the quick sharing of raw data, resulted in the efficient running of the project, which involved laboratories spread across various continents. A race to complete the genome sequence began when a private organisation was formed to sequence the human genome. In 2003 the sequence was completed, by both the private and publicly funded ventures. Commentators have said that this heralded a new age of medicine, with a 'genetic approach'.

Concerns about the exacerbation of inequities and inequalities in health because of progress in technology are not unfounded. Attempts to address these controversial issues relating to medical research might not be successful if the history of international (non-)engagement in dealing with medical and research ethics is anything to go by. The quest for autonomy in medical research is clearly an ongoing quest. The Nuremberg Code could have provided the starting point for international engagement with the ethics of medical research, but that was not the case. It nevertheless formed the basis for the establishment of international human rights law which is now being used to provide the framework for the international engagement with bioethics. Informed consent, the no-harm principle, benefit to society, and minimum standards for experimentation are some of the key statements in the Nuremberg Code and these principles can now be found in international and national law and policy. But at the time, the Code was considered 'a good code for barbarians, but an unnecessary code for ordinary physicians'. UNESCO, despite its mandate to promote scientific research for human welfare, did not get involved in the discussion on the doctors' trial and its judgement which incorporates the Code. The moral issue was seen to be the prevention of the State from interfering with research.

During the postwar years human experimentation greatly increased. Science and technology innovation was identified as a key driver to economic progress and prosperity in many countries. Simultaneously, and as a direct result of the encouragement of the growth of science and technology, the diversity of experiments in human experimentation increased and the nature of such experimentation also changed. Sophisticated technology involving mechanical, electronic and digital means now uses fragments of human material and has the potential to collect ever larger amounts of information. Such research is extremely international, with human samples and information crossing borders on an unprecedented scale. All this led to an innovation in scientific research by the end of the twentieth century – the integration of science with societal concerns. The ELSI programme covering research into ethical, legal and social issues as part of a scientific project (the HGP), was first launched in the US with a 3–5 per cent budget allocation from the scientific project. Its remit was to span a wide spectrum of ethical issues, such as fairness in the use of genetic information by employers and insurers, reproductive issues such as the use of genetic information for reproductive decision making, clinical issues about genetic testing and the concept of health, environmental issues, and the commercialisation of products. The main drawback of this programme and other similar ones is that it does not probe the ethics of doing 'big science' – the ethical concerns about inequalities in access to such research at the global level, etc. One very important ethical, legal and social concern is the concept of property rights in the human genome. This is contentious and complex at both the national and international level. While property rights that been granted on products derived from human tissue do not lie in the sample itself, and hence do not directly benefit the individual (the Moore case), rights granted to protect intellectual property (IP) derived from such samples benefit inventors and investors. This is a problematic situation if the tissue samples come from a certain indigenous group or population and no benefit accrues to them while the researcher, usually from a much more developed state, goes on to use the samples for new products to be backed by patent protection. However this is only one of the issues that could potentially exacerbate inequalities between the rich and the poor, between the healthy and the unhealthy, and between developed and developing country populations. There are a number of other issues that need to be considered and these are being addressed by various international organisations.

Bioethics, at the regulatory and governance level, has been supported by the use of human rights philosophy and law. In the US, the four-principle approach developed by Beauchamp and Childress included the principles of autonomy, non-malficience, beneficence and justice. Although it seems beneficial to use human rights jurisprudence to achieve bioethical goals and to concede that human rights has the potential to strengthen bioethical principles, it is argued that human rights law itself needs to be examined. Civil and political rights are much better protected than economic, social and cultural rights (ESCR). The emerging new technologies pose challenges both to first-generation and second-generation rights, but more so to second-generation rights. The central thesis of this book is that in order to address challenges posed by the discovery of the potential of the human genome to improve human health the legal governance of the human genome will only be able to function effectively if a unified theory of human rights is adopted. Such bioethical instruments underpinned by a unified theory will also be a major contribution to international human rights law. UNESCO, the UN General Assembly, the World Intellectual Property Organization (WIPO), the World Health Organization (WHO) and the European Community are contributing to this process. While UNESCO and its instruments have contributed the most to human rights through a sustained process of engagement on bioethics and human rights its instruments also list a number of new human rights, which add to the proliferation of human rights and endanger their fundamental nature. On the other hand, these rights can be construed as merely the applications of already established rights under the international system.

The status of the human genome is an important issue – is it a subject or an object of international law? Is it a unique entity? Does it have the status of a human being? Does it have special status because of its importance to the human person? Is it a special natural resource? The nature of the human genome is complex indeed. It contains the code that is required for the growth and development of every human being. However, it is not the sole contributor to human development. The genome is not a static entity; it is constantly shaping and being shaped by external factors. The likelihood of sharing common characteristics increases among close family members but an individual's genome is connected to others both within and outside the species. It could be construed both as a natural resource and as a special entity with a special status by itself. The debates at the International Bioethics Committee (IBC) reflect the complexity of the issue and have covered good ground in exploring the science and ethics of human genome governance.

The 1997 Universal Declaration on the Human Genome and Human Rights (adopted by the UN General Assembly in 1998) was the first instrument of its kind to elaborate rules for the governance of the human genome; it embodied a common heritage of mankind (CHM) framework and reiterated a number of rights including the right against discrimination and the right to health. The debate remains whether the right to informed consent is a bioethical principle or a human right. Freedom of thought and solidarity are other notable inclusions in the Universal Declaration, which is arguably part of 'soft law' in the area of human genome governance. Numerous national legislature and international instruments have incorporated principles found in the 1997 Universal Declaration, along with overt acceptance of it. The Economic and Social Council, a subsidiary body of the UN, addressed human rights and bioethics after the adoption of the 1997 Universal Declaration through one its bodies – the Sub-Commission on the Promotion and Protection of Human Rights. A Special Rapporteur was appointed in 2002 to address this issue and to report to the Sub-Commission. A human rights-based systematic approach to bioethical issues or the 1997 Universal Declaration is not evident in the Special Rapporteur's reports, although there is detailed analysis of discrimination based on genetic traits. They are interim reports that focus extensively on the breadth of the issues involved and draw some initial conclusions relating to intellectual property rights (IPR) and other issues. They indicate a moving away from a CHM regime on the grounds that such a regime would lead to a decrease if not a grinding to a halt of scientific research on the human genome. Such a conclusion is not unsubstantiated because deep seabed mining is seen to have been hampered because of the CHM regime. However, the report is based on the presumption that the human genome resource is a resource similar to deep seabed minerals and that a CHM regime would be applied to it without any modification. Given that extensive reform or even overhaul of the IP system relating to patents is not unforeseeable in the future, and that a reformed system would need to distribute property rights in an equitable manner, a CHM template is already at hand to provide a model. It needs to be adapted and modified to suit a special resource such as the human genome but it exists as a normative concept in international law.

WHO, within whose remit falls the promotion of the attainment of the right of all people to the highest standard of health, does not use a rights-based approach in the fulfilment of its mandate. It does not even favour a legal approach. The first ever treaty orchestrated by WHO on tobacco use has now entered into force and the focus on the individual is prominent therein, which however still avoids an outright rights-based approach. WHO's contribution to the governance of the human genome is the linkage it makes between trade, IP and health in the Tobacco Treaty and its resolutions on health. WHO is concerned about the impact of the disincentives the current innovation systems create in addressing health issues of low-return markets. It toys with, in one of its resolutions, the idea of an alternative regime of property and IP rights.

UNESCO, through its continuous and sustained engagement in the governance of the human genome has contributed enormously by bringing new issues on the agenda, by promoting debate on such issues and by providing guidelines for States to follow in regard to these issues. Through their activities the Human Rights Council and WHO have contributed to the governance of the human genome to a lesser extent but they have an important role to play in the future on property rights and health. All three organisations are contributing to the strengthening of second-generation rights and derive the basis of their frameworks from the same set of rights.

Plant genetic resources were declared the CHM in the early 1980s. The CHM concept was then modified after the adoption of the Convention on Biological Diversity in 1992, which declared such resources to fall within the sovereign power of States. So has the CHM been sidelined following its downgrading with respect to plant genetic resources? Is it really the case that the CHM model has no role in the international governance of genetic resources? It is evident from the 1997 Universal Declaration that the CHM concept plays more than a symbolic role. Even if the intention was to tone down the notion of the CHM in Article 1, other provisions of the Universal Declaration embody elements of a CHM regime. Rather more useful to human genome governance would be to question whether, in the light of the concerns about IP, health and human rights, it would be profitable to examine the different CHM models in international law before shelving the concept. The man aim of this book is to perform such an examination.

As is demonstrated by the negotiations during the creation of the 1982 Convention on the Law of the Sea and the 1979 Moon Treaty, States were grappling with precisely the same kinds of issues then as they are now with respect to the human genome. The overriding principle in the normative CHM models is one of regulation. The same is true of the biosciences; embryo research is tightly regulated in technologically advanced countries around the world. Research ethics committees sit on scientific research proposals to make sure they meet various criteria, both legal and ethical. The CHM is clearly not about preservation; it is a model that allows for exploitation of the resource in an equitable manner. Non-appropriation, international management and benefit sharing are the mechanisms through which equitable exploitation of the resource would be carried out. Alongside these three main elements, the CHM model promotes the peaceful use of a natural resource, in keeping with objectives of the UN Charter to maintain international peace and security. The 1997 Universal Declaration contains a viable outline for a CHM model of human genome governance, one that is ready to use as a subject of further discussion and debate. The 1997 Universal Declaration contains only an outline and is by no means complete. The benefit-sharing provisions, although onerous at first glance, are actually limited by the notion of human dignity, and of the right to development (RTD). The RTD is not an open-ended right, and in the context of the human genome, is centred on the right to health. The right to health, explored through Gewirthian human rights theory, can also be defined and identified as a right wherein the responsibility for the fulfilment of such a right is restricted to achievable goals.

Morality and ethics play a role in formulating international law. Human rights law is a case in point and Bioethics is a more recent example. Gewirth's rational human rights theory, based on the Principle of Generic Consistency (PGC), enables us to explore human dignity, human rights and human needs. It is especially useful in building a hierarchy of rights and in addressing the issue of responsibilities. In this theory value is placed on action. To be able to act and to act to fulfil an objective are two different and distinct stages. It can be argued that the inability of a person to be able to act is a violation of human dignity. In order to ensure that we respect human dignity, individuals have to be endowed with this basic ability to act so as to fulfil their chosen purposes. Gewirth argues that we would not be consistent in our views if we were not to accept that such ability is as important to us as it is to other persons. And if that is true then we have a duty to secure that ability for ourselves and, to the extent that it is possible, to secure it for others too. The ability to act is a basic need, and such basic needs can be fulfilled by a plurality of criteria – they are therefore described by Gewirth as 'basic generic needs'. These needs correspond to rights, which are basic generic rights. The relationship between individuals who are rationally consistent in their approach to recognising the importance of generic rights to themselves and others is the relationship of mutuality. Mutuality is characterised not by quid pro quo but by the recognition that one must do to the other anything that would help towards securing their generic needs. Such a conceptualisation of rights recognises positive rights and negative rights and incorporates a notion of solidarity based on the fulfilment of generic needs. It constructs a hierarchy of rights consisting of 'basic wellbeing' (fulfilment of basic generic needs) and 'freedom' (a condition which enables choice of action). Such a hierarchy reorganises the current normative hierarchy where civil and political rights take precedence over ECSR. It creates a different hierarchy based on generic needs. Solidarity involves a notion of responsibility which does not directly correlate to duties. Human rights are based not on strong atomistic individualism but on ethical individualism. Applying Gewirthian analysis to the elements of the CHM framework reveals a correspondence between the concepts in the theory and the CHM elements.

1 Non-appropriation – negative right.
2 International governance – positive right to assert freedom by taking part in decision making.
3 Peaceful uses – protection of basic human rights, such as the right to life.
4 Benefit sharing – Development of productive agency, a duty to cooperate (reciprocity).

The right to productive agency was devised by Gewirth to highlight the economic and social rights required to fulfil basic generic needs. Productive agency involves possessing the generic ability to act, along with the freedom to choose further goals which can be secured through such ability; productive agents are not dependent agents. The CHM framework is a regime that incorporates negative and positive rights, with the fundamental objective of securing to individuals their right to productive agency.

The right to productive agency is akin to the RTD in international law. The 1986 Declaration on the Right to Development and the 1993 Vienna Declaration and Programme of Action placed the RTD highest in the hierarchy of human rights. The RTD is defined in Article 1 of the 1986 Declaration as:

> an inalienable human right by virtue of which every person and all peoples are entitled to participate in, and contribute to, and enjoy economic, social, cultural and political development, in which all human rights and fundamental freedoms can be fully realised.

The operative part of this article – 'to participate in, and to contribute to' reflects the two stages of action indentified in Gewirthian theory. The ability to participate is the ability to act, and the chosen purpose of such action is a contribution to the enjoyment of human rights by all other persons. The RTD is not recognised explicitly in binding international law sources, as some other rights are. In order to implement the RTD, a Human Rights Commissioner was appointed in 1993 whose main mandate was to promote the RTD. In 2002 an Independent Expert was appointed to examine the implementation of this right and to provide recommendations for strengthening such implementation. He identified action on three issues to support the fulfilment of the RTD – food, health and primary education. He also recommended the formation of 'development compacts' in order to achieve minimum levels of human security. The Universal Declaration on Bioethics and Human Rights was adopted by the General Conference of UNESCO in 2005, and when this is read in conjunction with the CHM provisions of the 1997 Universal Declaration a framework within which health can be construed of in a similar context.

CHM frameworks are understood as preventing the grant of IPR in genetic research. This understanding is the mainstay of the opposition to the application of the CHM notion to the human genome. Normative models of CHM, such as in the Convention on the Law of the Sea, do not preclude grant of monopoly or exclusivity in the enjoyment of resources. They provide a property model different to the existing models of ownership and property rights. Such a property model sets out to reconcile monopoly interests with public interests and seeks to address the relationship between the effect of accumulation of property rights and the accruing of power from such accumulation of rights, which has an adverse effect on the fulfilment of rights of those who cannot participate in sharing such power. The Gewirthian framework lends itself to providing a reconceptualisation of property, and this has been formulated by Beyleveld and Brownsword as their 'rule-preclusionary' theory of property. This theory provides a basis for addressing property rights as a human right in order to address current issues in genomic research and also to provide an understanding of property that could be incorporated into a CHM model. Traditionally property rights are conceived of as a bundle of rights that provide greater control to some people over things than others. The rule-preclusionary principle incorporates this idea of the exclusion of others but also attaches the condition that such preclusion of others should be required in order to secure for oneself generic rights of agency. So the ownership of the object only extends to the extent that it is required by the individual for their generic needs and any other rights-control over the intangible or tangible object can only be had so far as it does not adversely affect others' generic rights.

Property rights and the economics of genomic research are vital issues that have an impact on innovation, health, human rights and the economic advancement of societies and States. This book has demonstrated that these are issues that have been addressed by the international community in the past decades, and that useful concepts and mechanisms can be found in international law to address problematic aspects of the governance of the human genome. Such debates contribute to the formation of international law and advance adherence to the rule of law at the international level.

Appendix

Universal Declaration on the Human Genome and Human Rights

11 November 1997

The General Conference

Recalling that the Preamble of UNESCO's Constitution refers to 'the democratic principles of the dignity, equality and mutual respect of men', rejects any 'doctrine of the inequality of men and races', stipulates 'that the wide diffusion of culture, and the education of humanity for justice and liberty and peace are indispensable to the dignity of men and constitute a sacred duty which all the nations must fulfil in a spirit of mutual assistance and concern', proclaims that 'peace must be founded upon the intellectual and moral solidarity of mankind', and states that the Organization seeks to advance, 'through the educational and scientific and cultural relations of the peoples of the world, the objectives of international peace and of the common welfare of mankind for which the United Nations Organization was established and which its Charter proclaims'

Solemnly recalling its attachment to the universal principles of human rights, affirmed in particular in the Universal Declaration of Human Rights of 10 December 1948 and in the two International United Nations Covenants on Economic, Social and Cultural Rights and on Civil and Political Rights of 16 December 1966, in the United Nations Convention on the Prevention and Punishment of the Crime of Genocide of 9 December 1948, the International United Nations Convention on the Elimination of All Forms of Racial Discrimination of 21 December 1965, the United Nations Declaration on the Rights of Mentally Retarded Persons of 20 December 1971, the United Nations Declaration on the Rights of Disabled Persons of 9 December 1975, the United Nations Convention on the Elimination of All Forms of Discrimination Against Women of 18 December 1979, the United Nations Declaration of Basic Principles of Justice for Victims of Crime and Abuse of Power of 29 November 1985, the United Nations Convention on the Rights of the Child of 20 November 1989, the United Nations Standard Rules on the Equalization of Opportunities for Persons with Disabilities of 20 December 1993, the Convention on the Prohibition of the Development, Production and Stockpiling of Bacteriological (Biological) and Toxin Weapons and on their Destruction of 16 December 1971, the UNESCO Convention against Discrimination in Education of 14 December 1960, the UNESCO Declaration of the Principles of International Cultural Co-operation of 4 November 1966, the UNESCO Recommendation on the Status of Scientific Researchers of 20 November 1974, the UNESCO Declaration on Race and Racial Prejudice of 27 November 1978, the ILO Convention (No. 111) concerning Discrimination in Respect of Employment and Occupation of 25 June 1958 and the ILO Convention (No. 169) concerning Indigenous and Tribal Peoples in Independent Countries of 27 June 1989

Bearing in mind, and without prejudice to, the international instruments which could have a bearing on the applications of genetics in the field of intellectual property, inter alia the Berne Convention for the Protection of Literary and Artistic Works of 9 September 1886 and the UNESCO Universal Copyright Convention of 6 September 1952, as last revised at Paris on 24 July 1971, the Paris Convention for the Protection of Industrial Property of 20 March 1883, as last revised at Stockholm on 14 July 1967, the Budapest Treaty of the WIPO on International Recognition of the Deposit of Micro-organisms for the Purposes of Patent Procedures of 28 April 1977, and the Trade Related Aspects of Intellectual Property Rights Agreement (TRIPs) annexed to the Agreement establishing the World Trade Organization, which entered into force on 1 January 1995

Bearing in mind also the United Nations Convention on Biological Diversity of 5 June 1992 and emphasizing in that connection that the recognition of the genetic diversity of humanity must not give rise to any interpretation of a social or political nature which could call into question 'the inherent dignity and (…) the equal and inalienable rights of all members of the human family', in accordance with the Preamble to the Universal Declaration of Human Rights

Recalling 22 C/Resolution 13.1, 23 C/Resolution 13.1, 24 C/Resolution 13.1, 25 C/Resolutions 5.2 and 7.3, 27 C/Resolution 5.15 and 28 C/Resolutions 0.12, 2.1 and 2.2, urging UNESCO to promote and develop ethical studies, and the actions arising out of them, on the consequences of scientific and technological progress in the fields of biology and genetics, within the framework of respect for human rights and fundamental freedoms

Recognizing that research on the human genome and the resulting applications open up vast prospects for progress in improving the health of individuals and of humankind as a whole, but **emphasizing** that such research should fully respect human dignity, freedom and human rights, as well as the prohibition of all forms of discrimination based on genetic characteristics

Proclaims the principles that follow and **adopts** the present Declaration.

A Human dignity and the human genome

Article 1

The human genome underlies the fundamental unity of all members of the human family, as well as the recognition of their inherent dignity and diversity. In a symbolic sense, it is the heritage of humanity.

Article 2

(a) Everyone has a right to respect for their dignity and for their rights regardless of their genetic characteristics.
(b) That dignity makes it imperative not to reduce individuals to their genetic characteristics and to respect their uniqueness and diversity.

Article 3

The human genome, which by its nature evolves, is subject to mutations. It contains potentialities that are expressed differently according to each individual's natural and social environment, including the individual's state of health, living conditions, nutrition and education.

Article 4

The human genome in its natural state shall not give rise to financial gains.

B Rights of the persons concerned

Article 5

(a) Research, treatment or diagnosis affecting an individual's genome shall be undertaken only after rigorous and prior assessment of the potential risks and benefits pertaining thereto and in accordance with any other requirement of national law.
(b) In all cases, the prior, free and informed consent of the person concerned shall be obtained. If the latter is not in a position to consent, consent or authorization shall be obtained in the manner prescribed by law, guided by the person's best interest.
(c) The right of each individual to decide whether or not to be informed of the results of genetic examination and the resulting consequences should be respected.
(d) In the case of research, protocols shall, in addition, be submitted for prior review in accordance with relevant national and international research standards or guidelines.

(e) If according to the law a person does not have the capacity to consent, research affecting his or her genome may only be carried out for his or her direct health benefit, subject to the authorization and the protective conditions prescribed by law. Research which does not have an expected direct health benefit may only be undertaken by way of exception, with the utmost restraint, exposing the person only to a minimal risk and minimal burden and if the research is intended to contribute to the health benefit of other persons in the same age category or with the same genetic condition, subject to the conditions prescribed by law, and provided such research is compatible with the protection of the individual's human rights.

Article 6

No one shall be subjected to discrimination based on genetic characteristics that is intended to infringe or has the effect of infringing human rights, fundamental freedoms and human dignity.

Article 7

Genetic data associated with an identifiable person and stored or processed for the purposes of research or any other purpose must be held confidential in the conditions set by law.

Article 8

Every individual shall have the right, according to international and national law, to just reparation for any damage sustained as a direct and determining result of an intervention affecting his or her genome.

Article 9

In order to protect human rights and fundamental freedoms, limitations to the principles of consent and confidentiality may only be prescribed by law, for compelling reasons within the bounds of public international law and the international law of human rights.

C Research on the human genome Article 10

No research or research applications concerning the human genome, in particular in the fields of biology, genetics and medicine, should prevail over respect for the human rights, fundamental freedoms and human dignity of individuals or, where applicable, of groups of people.

Article 11

Practices which are contrary to human dignity, such as reproductive cloning of human beings, shall not be permitted. States and competent international organizations are invited to co-operate in identifying such practices and in taking, at national or international level, the measures necessary to ensure that the principles set out in this Declaration are respected.

Article 12

(a) Benefits from advances in biology, genetics and medicine, concerning the human genome, shall be made available to all, with due regard for the dignity and human rights of each individual.
(b) Freedom of research, which is necessary for the progress of knowledge, is part of freedom of thought. The applications of research, including applications in biology, genetics and medicine, concerning the human genome, shall seek to offer relief from suffering and improve the health of individuals and humankind as a whole.

D Conditions for the exercise of scientific activity Article 13

The responsibilities inherent in the activities of researchers, including meticulousness, caution, intellectual honesty and integrity in carrying out their research as well as in the presentation and utilization of their findings, should be the subject of particular attention in the framework of research on the human genome, because of its ethical and social implications. Public and private science policy-makers also have particular responsibilities in this respect.

Article 14

States should take appropriate measures to foster the intellectual and material conditions favourable to freedom in the conduct of research on the human genome and to consider the ethical, legal, social and economic implications of such research, on the basis of the principles set out in this Declaration.

Article 15

States should take appropriate steps to provide the framework for the free exercise of Research on the human genome with due regard for the principles set out in this Declaration, in order to safeguard respect for human rights, fundamental freedoms and human dignity and to protect public health. They should seek to ensure that research results are not used for non-peaceful purposes.

Article 16

States should recognize the value of promoting, at various levels, as appropriate, the establishment of independent, multidisciplinary and pluralist ethics committees to assess the ethical, legal and social issues raised by research on the human genome and its applications.

E Solidarity and international co-operation Article 17

States should respect and promote the practice of solidarity towards individuals, families and population groups who are particularly vulnerable to or affected by disease or disability of a genetic character. They should foster, inter alia, research on the identification, prevention and treatment of genetically based and genetically influenced diseases, in particular rare as well as endemic diseases which affect large numbers of the world's population.

Article 18

States should make every effort, with due and appropriate regard for the principles set out in this Declaration, to continue fostering the international dissemination of scientific knowledge concerning the human genome, human diversity and genetic research and, in that regard, to foster scientific and cultural co-operation, particularly between industrialized and developing countries.

Article 19

(a) In the framework of international co-operation with developing countries, states should seek to encourage measures enabling:

- (i) assessment of the risks and benefits pertaining to research on the human genome to be carried out and abuse to be prevented;
- (ii) the capacity of developing countries to carry out research on human biology and genetics, taking into consideration their specific problems, to be developed and strengthened;
- (iii) developing countries to benefit from the achievements of scientific and technological research so that their use in favour of economic and social progress can be to the benefit of all;
- (iv) the free exchange of scientific knowledge and information in the areas of biology, genetics and medicine to be promoted.

(b) Relevant international organizations should support and promote the initiatives taken by states for the above-mentioned purposes.

F Promotion of the principles set out in the Declaration

Article 20

States should take appropriate measures to promote the principles set out in the Declaration, through education and relevant means, inter alia through the conduct of research and training in interdisciplinary fields and through the promotion of education in bioethics, at all levels, in particular for those responsible for science policies.

Article 21

States should take appropriate measures to encourage other forms of research, training and information dissemination conducive to raising the awareness of society and all of its members of their responsibilities regarding the fundamental issues relating to the defence of human dignity which may be raised by research in biology, in genetics and in medicine, and its applications. They should also undertake to facilitate on this subject an open international discussion, ensuring the free expression of various sociocultural, religious and philosophical opinions.

G Implementation of the Declaration

Article 22

States should make every effort to promote the principles set out in this Declaration and should, by means of all appropriate measures, promote their implementation.

Article 23

States should take appropriate measures to promote, through education, training and information dissemination, respect for the above-mentioned principles and to foster their recognition and effective application. States should also encourage exchanges and networks among independent ethics committees, as they are established, to foster full collaboration.

Article 24

The International Bioethics Committee of UNESCO should contribute to the dissemination of the principles set out in this Declaration and to the further examination of issues raised by their applications and by the evolution of the technologies in question. It should organize appropriate consultations with parties concerned, such as vulnerable groups. It should make recommendations, in accordance with UNESCO's statutory procedures, addressed to the General Conference and give advice concerning the follow-up of this Declaration, in particular regarding the identification of practices that could be contrary to human dignity, such as germ-line interventions.

Article 25

Nothing in this Declaration may be interpreted as implying for any state, group or person any claim to engage in any activity or to perform any act contrary to human rights and fundamental freedoms, including the principles set out in this Declaration.

International Declaration on Human Genetic Data

16 October 2003

The General Conference

Recalling the Universal Declaration of Human Rights of 10 December 1948, the two United Nations International Covenants on Economic, Social and Cultural Rights and on Civil and Political Rights of 16 December 1966, the United Nations International Convention on the Elimination of All Forms of Racial Discrimination of 21 December 1965, the United Nations Convention on the Elimination of All Forms of Discrimination against Women of 18 December 1979, the United Nations Convention on the Rights of the Child of 20 November 1989, the United Nations Economic and Social Council resolutions 2001/39 on Genetic Privacy and Non-Discrimination of 26 July 2001 and 2003/232 on Genetic Privacy and Non-Discrimination of 22 July 2003, the ILO Convention (No. 111) concerning Discrimination in Respect of Employment and Occupation of 25 June 1958, the UNESCO Universal Declaration on Cultural Diversity of 2 November 2001, the Trade Related Aspects of Intellectual Property Rights Agreement (TRIPS) annexed to the Agreement establishing the World Trade Organization, which entered into force on 1 January 1995, the Doha Declaration on the TRIPS Agreement and Public Health of 14 November 2001 and the other international human rights instruments adopted by the United Nations and the specialized agencies of the United Nations system

Recalling more particularly the Universal Declaration on the Human Genome and Human Rights which it adopted, unanimously and by acclamation, on 11 November 1997 and which was endorsed by the United Nations General Assembly on 9 December 1998 and the Guidelines for the implementation of the Universal Declaration on the Human Genome and Human Rights which it endorsed on 16 November 1999 by 30 C/Resolution 23

Welcoming the broad public interest worldwide in the Universal Declaration on the Human Genome and Human Rights, the firm support it has received from the international community and its impact in Member States drawing upon it for their legislation, regulations, norms and standards, and ethical codes of conduct and guidelines

Bearing in mind the international and regional instruments, national laws, regulations and ethical texts relating to the protection of human rights and fundamental freedoms and to respect for human dignity as regards the collection, processing, use and storage of scientific data, as well as of medical data and personal data

Recognizing that genetic information is part of the overall spectrum of medical data and that the information content of any medical data, including genetic data and proteomic data, is highly contextual and dependent on the particular circumstances

Also recognizing that human genetic data have a special status on account of their sensitive nature since they can be predictive of genetic predispositions concerning individuals and that the power of predictability can be stronger than assessed at the time of deriving the data; they may have a significant impact on the family, including offspring, extending over generations, and in some instances on the whole group; they may contain information the significance of which is not necessarily known at the time of the collection of biological samples; and they may have cultural significance for persons or groups

Emphasizing that all medical data, including genetic data and proteomic data, regardless of their apparent information content, should be treated with the same high standards of confidentiality

Noting the increasing importance of human genetic data for economic and commercial purposes

Having regard to the special needs and vulnerabilities of developing countries and the need to reinforce international cooperation in the field of human genetics

Considering that the collection, processing, use and storage of human genetic data are of paramount importance for the progress of life sciences and medicine, for their applications and for the use of such data for non-medical purposes

Also considering that the growing amount of personal data collected makes genuine irretrievability increasingly difficult

Aware that the collection, processing, use and storage of human genetic data have potential risks for the exercise and observance of human rights and fundamental freedoms and respect for human dignity

Noting that the interests and welfare of the individual should have priority over the rights and interests of society and research

Reaffirming the principles established in the Universal Declaration on the Human Genome and Human Rights and the principles of equality, justice, solidarity and responsibility as well as respect for human dignity, human rights and fundamental freedoms, particularly freedom of thought and expression, including freedom of research, and privacy and security of the person, which must underlie the collection, processing, use and storage of human genetic data

Proclaims the principles that follow and **adopts** the present Declaration.

A General provisions

Article 1 – Aims and scope

(a) The aims of this Declaration are: to ensure the respect of human dignity and protection of human rights and fundamental freedoms in the collection, processing, use and storage of human genetic data, human proteomic data and of the biological samples from which they are derived, referred to hereinafter as "biological samples", in keeping with the requirements of equality, justice and solidarity, while giving due consideration to freedom of thought and expression, including freedom of research; to set out the principles which should guide States in the formulation of their legislation and their policies on these issues; and to form the basis for guidelines of good practices in these areas for the institutions and individuals concerned.
(b) Any collection, processing, use and storage of human genetic data, human proteomic data and biological samples shall be consistent with the international law of human rights.
(c) The provisions of this Declaration apply to the collection, processing, use and storage of human genetic data, human proteomic data and biological samples, except in the investigation, detection and prosecution of criminal offences and in parentage testing that are subject to domestic law that is consistent with the international law of human rights.

Article 2 – Use of terms

For the purposes of this Declaration, the terms used have the following meanings:

(i) Human genetic data: Information about heritable characteristics of individuals obtained by analysis of nucleic acids or by other scientific analysis;
(ii) Human proteomic data: Information pertaining to an individual's proteins including their expression, modification and interaction;
(iii) Consent: Any freely given specific, informed and express agreement of an individual to his or her genetic data being collected, processed, used and stored;
(iv) Biological samples: Any sample of biological material (for example blood, skin and bone cells or blood plasma) in which nucleic acids are present and which contains the characteristic genetic make-up of an individual;

(v) Population-based genetic study: A study which aims at understanding the nature and extent of genetic variation among a population or individuals within a group or between individuals across different groups;
(vi) Behavioural genetic study: A study that aims at establishing possible connections between genetic characteristics and behaviour;
(vii) Invasive procedure: Biological sampling using a method involving intrusion into the human body, such as obtaining a blood sample by using a needle and syringe;
(viii) Non-invasive procedure: Biological sampling using a method which does not involve intrusion into the human body, such as oral smears;
(ix) Data linked to an identifiable person: Data that contain information, such as name, birth date and address, by which the person from whom the data were derived can be identified;
(x) Data unlinked to an identifiable person: Data that are not linked to an identifiable person, through the replacement of, or separation from, all identifying information about that person by use of a code;
(xi) Data irretrievably unlinked to an identifiable person: Data that cannot be linked to an identifiable person, through destruction of the link to any identifying information about the person who provided the sample;
(xii) Genetic testing: A procedure to detect the presence or absence of, or change in, a particular gene or chromosome, including an indirect test for a gene product or other specific metabolite that is primarily indicative of a specific genetic change;
(xiii) Genetic screening: Large-scale systematic genetic testing offered in a programme to a population or subsection thereof intended to detect genetic characteristics in asymptomatic people;
(xiv) Genetic counselling: A procedure to explain the possible implications of the findings of genetic testing or screening, its advantages and risks and where applicable to assist the individual in the long-term handling of the consequences; It takes place before and after genetic testing and screening;
(xv) Cross-matching: Matching of information about an individual or a group contained in various data files set up for different purposes.

Article 3 – Person's identity

Each individual has a characteristic genetic make-up. Nevertheless, a person's identity should not be reduced to genetic characteristics, since it involves complex educational, environmental and personal factors and emotional, social, spiritual and cultural bonds with others and implies a dimension of freedom.

Article 4 – Special status

(a) Human genetic data have a special status because:

(i) they can be predictive of genetic predispositions concerning individuals;

(ii) they may have a significant impact on the family, including offspring, extending over generations, and in some instances on the whole group to which the person concerned belongs;
(iii) they may contain information the significance of which is not necessarily known at the time of the collection of the biological samples;
(iv) they may have cultural significance for persons or groups.

(b) Due consideration should be given to the sensitivity of human genetic data and an appropriate level of protection for these data and biological samples should be established.

Article 5 – Purposes

Human genetic data and human proteomic data may be collected, processed, used and stored only for the purposes of:

i diagnosis and health care, including screening and predictive testing;
ii medical and other scientific research, including epidemiological, especially population-based genetic studies, as well as anthropological or archaeological studies, collectively referred to hereinafter as "medical and scientific research";
iii forensic medicine and civil, criminal and other legal proceedings, taking into account the provisions of Article 1(c);
iv or any other purpose consistent with the Universal Declaration on the Human Genome and Human Rights and the international law of human rights.

Article 6 – Procedures

(a) It is ethically imperative that human genetic data and human proteomic data be collected, processed, used and stored on the basis of transparent and ethically acceptable procedures. States should endeavour to involve society at large in the decision-making process concerning broad policies for the collection, processing, use and storage of human genetic data and human proteomic data and the evaluation of their management, in particular in the case of population-based genetic studies. This decision-making process, which may benefit from international experience, should ensure the free expression of various viewpoints.
(b) Independent, multidisciplinary and pluralist ethics committees should be promoted and established at national, regional, local or institutional levels, in accordance with the provisions of Article 16 of the Universal Declaration on the Human Genome and Human Rights. Where appropriate, ethics committees at national level should be consulted with regard to the establishment of standards, regulations and guidelines for the collection, processing, use and storage of human genetic data, human proteomic data

and biological samples. They should also be consulted concerning matters where there is no domestic law. Ethics committees at institutional or local levels should be consulted with regard to their application to specific research projects.

(c) When the collection, processing, use and storage of human genetic data, human proteomic data or biological samples are carried out in two or more States, the ethics committees in the States concerned, where appropriate, should be consulted and the review of these questions at the appropriate level should be based on the principles set out in this Declaration and on the ethical and legal standards adopted by the States concerned.

(d) It is ethically imperative that clear, balanced, adequate and appropriate information shall be provided to the person whose prior, free, informed and express consent is sought. Such information shall, alongside with providing other necessary details, specify the purpose for which human genetic data and human proteomic data are being derived from biological samples, and are used and stored. This information should indicate, if necessary, risks and consequences. This information should also indicate that the person concerned can withdraw his or her consent, without coercion, and this should entail neither a disadvantage nor a penalty for the person concerned.

Article 7 – Non-discrimination and non-stigmatization

(a) Every effort should be made to ensure that human genetic data and human proteomic data are not used for purposes that discriminate in a way that is intended to infringe, or has the effect of infringing human rights, fundamental freedoms or human dignity of an individual or for purposes that lead to the stigmatization of an individual, a family, a group or communities.

(b) In this regard, appropriate attention should be paid to the findings of population-based genetic studies and behavioural genetic studies and their interpretations.

B Collection

Article 8 – Consent

(a) Prior, free, informed and express consent, without inducement by financial or other personal gain, should be obtained for the collection of human genetic data, human proteomic data or biological samples, whether through invasive or non-invasive procedures, and for their subsequent processing, use and storage, whether carried out by public or private institutions. Limitations on this principle of consent should only be prescribed for compelling reasons by domestic law consistent with the international law of human rights.

(b) When, in accordance with domestic law, a person is incapable of giving informed consent, authorization should be obtained from the legal representative, in accordance with domestic law. The legal representative should have regard to the best interest of the person concerned.

(c) An adult not able to consent should as far as possible take part in the authorization procedure. The opinion of a minor should be taken into consideration as an increasingly determining factor in proportion to age and degree of maturity.

(d) In diagnosis and health care, genetic screening and testing of minors and adults not able to consent will normally only be ethically acceptable when they have important implications for the health of the person and have regard to his or her best interest.

Article 9 – Withdrawal of consent

(a) When human genetic data, human proteomic data or biological samples are collected for medical and scientific research purposes, consent may be withdrawn by the person concerned unless such data are irretrievably unlinked to an identifiable person. In accordance with the provisions of Article 6(d), withdrawal of consent should entail neither a disadvantage nor a penalty for the person concerned.

(b) When a person withdraws consent, the person's genetic data, proteomic data and biological samples should no longer be used unless they are irretrievably unlinked to the person concerned.

(c) If not irretrievably unlinked, the data and biological samples should be dealt with in accordance with the wishes of the person. If the person's wishes cannot be determined or are not feasible or are unsafe, the data and biological samples should either be irretrievably unlinked or destroyed.

Article 10 – The right to decide whether or not to be informed about research results

When human genetic data, human proteomic data or biological samples are collected for medical and scientific research purposes, the information provided at the time of consent should indicate that the person concerned has the right to decide whether or not to be informed of the results. This does not apply to research on data irretrievably unlinked to identifiable persons or to data that do not lead to individual findings concerning the persons who have participated in such a research. Where appropriate, the right not to be informed should be extended to identified relatives who may be affected by the results.

Article 11 – Genetic counselling

It is ethically imperative that when genetic testing that may have significant implications for a person's health is being considered, genetic counselling should be made available in an appropriate manner. Genetic counselling should be non-directive, culturally adapted and consistent with the best interest of the person concerned.

Article 12 – Collection of biological samples for forensic medicine or in civil, criminal and other legal proceedings

When human genetic data or human proteomic data are collected for the purposes of forensic medicine or in civil, criminal and other legal proceedings, including parentage testing, the collection of biological samples, in vivo or post-mortem, should be made only in accordance with domestic law consistent with the international law of human rights.

C Processing

Article 13 – Access

No one should be denied access to his or her own genetic data or proteomic data unless such data are irretrievably unlinked to that person as the identifiable source or unless domestic law limits such access in the interest of public health, public order or national security.

Article 14 – Privacy and confidentiality

(a) States should endeavour to protect the privacy of individuals and the confidentiality of human genetic data linked to an identifiable person, family or, where appropriate, group, in accordance with domestic law consistent with the international law of human rights.

(b) Human genetic data, human proteomic data and biological samples linked to an identifiable person should not be disclosed or made accessible to third parties, in particular, employers, insurance companies, educational institutions and the family, except for an important public interest reason in cases restrictively provided for by domestic law consistent with the international law of human rights or where the prior, free, informed and express consent of the person concerned has been obtained provided that such consent is in accordance with domestic law and the international law of human rights. The privacy of an individual participating in a study using human genetic data, human proteomic data or biological samples should be protected and the data should be treated as confidential.

(c) Human genetic data, human proteomic data and biological samples collected for the purposes of scientific research should not normally be linked to an identifiable person. Even when such data or biological samples are unlinked to an identifiable person, the necessary precautions should be taken to ensure the security of the data or biological samples.

(d) Human genetic data, human proteomic data and biological samples collected for medical and scientific research purposes can remain linked to an identifiable person, only if necessary to carry out the research and provided that the privacy of the individual and the confidentiality of the data or biological samples concerned are protected in accordance with domestic law.

(e) Human genetic data and human proteomic data should not be kept in a form which allows the data subject to be identified for any longer than is necessary for achieving the purposes for which they were collected or subsequently processed.

Article 15 – Accuracy, reliability, quality and security

The persons and entities responsible for the processing of human genetic data, human proteomic data and biological samples should take the necessary measures to ensure the accuracy, reliability, quality and security of these data and the processing of biological samples. They should exercise rigour, caution, honesty and integrity in the processing and interpretation of human genetic data, human proteomic data or biological samples, in view of their ethical, legal and social implications.

D Use

Article 16 – Change of purpose

(a) Human genetic data, human proteomic data and the biological samples collected for one of the purposes set out in Article 5 should not be used for a different purpose that is incompatible with the original consent, unless the prior, free, informed and express consent of the person concerned is obtained according to the provisions of Article 8(a) or unless the proposed use, decided by domestic law, corresponds to an important public interest reason and is consistent with the international law of human rights. If the person concerned lacks the capacity to consent, the provisions of Article 8(b) and (c) should apply mutatis mutandis.

(b) When prior, free, informed and express consent cannot be obtained or in the case of data irretrievably unlinked to an identifiable person, human genetic data may be used in accordance with domestic law or following the consultation procedures set out in Article 6(b).

Article 17 – Stored biological samples

(a) Stored biological samples collected for purposes other than set out in Article 5 may be used to produce human genetic data or human proteomic data with the prior, free, informed and express consent of the person concerned. However, domestic law may provide that if such data have significance for medical and scientific research purposes e.g. epidemiological studies, or public health purposes, they may be used for those purposes, following the consultation procedures set out in Article 6(b).

(b) The provisions of Article 12 should apply mutatis mutandis to stored biological samples used to produce human genetic data for forensic medicine.

Article 18 – Circulation and international cooperation

(a) States should regulate, in accordance with their domestic law and international agreements, the cross-border flow of human genetic data, human proteomic data and biological samples so as to foster international medical and scientific cooperation and ensure fair access to these data. Such a system should seek to ensure that the receiving party provides adequate protection in accordance with the principles set out in this Declaration.

(b) States should make every effort, with due and appropriate regard for the principles set out in this Declaration, to continue fostering the international dissemination of scientific knowledge concerning human genetic data and human proteomic data and, in that regard, to foster scientific and cultural cooperation, particularly between industrialized and developing countries.

(c) Researchers should endeavour to establish cooperative relationships, based on mutual respect with regard to scientific and ethical matters and, subject to the provisions of Article 14, should encourage the free circulation of human genetic data and human proteomic data in order to foster the sharing of scientific knowledge, provided that the principles set out in this Declaration are observed by the parties concerned. To this end, they should also endeavour to publish in due course the results of their research.

Article 19 – Sharing of benefits

(a) In accordance with domestic law or policy and international agreements, benefits resulting from the use of human genetic data, human proteomic data or biological samples collected for medical and scientific research should be shared with the society as a whole and the international community. In giving effect to this principle, benefits may take any of the following forms:

   (i) special assistance to the persons and groups that have taken part in the research;
   (ii) access to medical care;
   (iii) provision of new diagnostics, facilities for new treatments or drugs stemming from the research;
   (iv) support for health services;
   (v) capacity-building facilities for research purposes;
   (vi) development and strengthening of the capacity of developing countries to collect and process human genetic data, taking into consideration their specific problems;
   (vii) any other form consistent with the principles set out in this Declaration.

(b) Limitations in this respect could be provided by domestic law and international agreements.

E Storage

Article 20 – Monitoring and management framework

States may consider establishing a framework for the monitoring and management of human genetic data, human proteomic data and biological samples based on the principles of independence, multidisciplinarity, pluralism and transparency as well as the principles set out in this Declaration. This framework could also deal with the nature and purposes of the storage of these data.

Article 21 – Destruction

(a) The provisions of Article 9 apply mutatis mutandis in the case of stored human genetic data, human proteomic data and biological samples.
(b) Human genetic data, human proteomic data and the biological samples collected from a suspect in the course of a criminal investigation should be destroyed when they are no longer necessary, unless otherwise provided for by domestic law consistent with the international law of human rights.
(c) Human genetic data, human proteomic data and biological samples should be available for forensic purposes and civil proceedings only for as long as they are necessary for those proceedings, unless otherwise provided for by domestic law consistent with the international law of human rights.

Article 22 – Cross-matching

Consent should be essential for the cross-matching of human genetic data, human proteomic data or biological samples stored for diagnostic and health care purposes and for medical and other scientific research purposes, unless otherwise provided for by domestic law for compelling reasons and consistent with the international law of human rights.

F Promotion and implementation

Article 23 – Implementation

(a) States should take all appropriate measures, whether of a legislative, administrative or other character, to give effect to the principles set out in this Declaration, in accordance with the international law of human rights. Such measures should be supported by action in the sphere of education, training and public information.
(b) In the framework of international cooperation, States should endeavour to enter into bilateral and multilateral agreements enabling developing countries to build up their capacity to participate in generating and sharing scientific knowledge concerning human genetic data and the related know-how.

Article 24 – Ethics education, training and information

In order to promote the principles set out in this Declaration, States should endeavour to foster all forms of ethics education and training at all levels as well as to encourage information and knowledge dissemination programmes about human genetic data. These measures should aim at specific audiences, in particular researchers and members of ethics committees, or be addressed to the public at large. In this regard, States should encourage the participation of international and regional intergovernmental organizations and international, regional and national non-governmental organizations in this endeavour.

Article 25 – Roles of the International Bioethics Committee (IBC) and the Intergovernmental Bioethics Committee (IGBC)

The International Bioethics Committee (IBC) and the Intergovernmental Bioethics Committee (IGBC) shall contribute to the implementation of this Declaration and the dissemination of the principles set out therein. On a collaborative basis, the two Committees should be responsible for its monitoring and for the evaluation of its implementation, inter alia, on the basis of reports provided by States. The two Committees should be responsible in particular for the formulation of any opinion or proposal likely to further the effectiveness of this Declaration. They should make recommendations in accordance with UNESCO's statutory procedures, addressed to the General Conference.

Article 26 – Follow-up action by UNESCO

UNESCO shall take appropriate action to follow up this Declaration so as to foster progress of the life sciences and their applications through technologies, based on respect for human dignity and the exercise and observance of human rights and fundamental freedoms.

Article 27 – Denial of acts contrary to human rights, fundamental freedoms and human dignity

Nothing in this Declaration may be interpreted as implying for any State, group or person any claim to engage in any activity or to perform any act contrary to human rights, fundamental freedoms and human dignity, including, in particular, the principles set out in this Declaration.

Universal Declaration on Bioethics and Human Rights

19 October 2005

The General Conference

Conscious of the unique capacity of human beings to reflect upon their own existence and on their environment, to perceive injustice, to avoid danger, to assume responsibility, to seek cooperation and to exhibit the moral sense that gives expression to ethical principles

Reflecting on the rapid developments in science and technology, which increasingly affect our understanding of life and life itself, resulting in a strong demand for a global response to the ethical implications of such developments

Recognizing that ethical issues raised by the rapid advances in science and their technological applications should be examined with due respect to the dignity of the human person and universal respect for, and observance of, human rights and fundamental freedoms

Resolving that it is necessary and timely for the international community to state universal principles that will provide a foundation for humanity's response to the ever-increasing dilemmas and controversies that science and technology present for humankind and for the environment

Recalling the Universal Declaration of Human Rights of 10 December 1948, the Universal Declaration on the Human Genome and Human Rights adopted by the General Conference of UNESCO on 11 November 1997 and the International Declaration on Human Genetic Data adopted by the General Conference of UNESCO on 16 October 2003

Noting the United Nations International Covenant on Economic, Social and Cultural Rights and the International Covenant on Civil and Political Rights of 16 December 1966, the United Nations International Convention on the Elimination of All Forms of Racial Discrimination of 21 December 1965, the United Nations Convention on the Elimination of All Forms of Discrimination against Women of 18 December 1979, the United Nations Convention on the Rights of the Child of 20 November 1989, the United Nations Convention on Biological Diversity of 5 June 1992, the Standard Rules on the Equalization of Opportunities for Persons with Disabilities adopted by the General Assembly of the United Nations in 1993, the UNESCO Recommendation on the Status of Scientific Researchers of 20 November 1974, the UNESCO Declaration on Race and Racial Prejudice of 27 November 1978, the UNESCO Declaration on the Responsibilities of the

Present Generations Towards Future Generations of 12 November 1997, the UNESCO Universal Declaration on Cultural Diversity of 2 November 2001, the ILO Convention 169 concerning Indigenous and Tribal Peoples in Independent Countries of 27 June 1989, the International Treaty on Plant Genetic Resources for Food and Agriculture which was adopted by the FAO Conference on 3 November 2001 and entered into force on 29 June 2004, the Agreement on Trade-Related Aspects of Intellectual Property Rights (TRIPS) annexed to the Marrakech Agreement establishing the World Trade Organization, which entered into force on 1 January 1995, the Doha Declaration on the TRIPS Agreement and Public Health of 14 November 2001 and other relevant international instruments adopted by the United Nations and the specialized agencies of the United Nations system, in particular the Food and Agriculture Organization of the United Nations (FAO) and the World Health Organization (WHO)

Also noting international and regional instruments in the field of bioethics, including the Convention for the Protection of Human Rights and Dignity of the Human Being with regard to the Application of Biology and Medicine: Convention on Human Rights and Biomedicine of the Council of Europe, which was adopted in 1997 and entered into force in 1999, together with its Additional Protocols, as well as national legislation and regulations in the field of bioethics and the international and regional codes of conduct and guidelines and other texts in the field of bioethics, such as the Declaration of Helsinki of the World Medical Association on Ethical Principles for Medical Research Involving Human Subjects, adopted in 1964 and amended in 1975, 1983, 1989, 1996 and 2000 and the International Ethical Guidelines for Biomedical Research Involving Human Subjects of the Council for International Organizations of Medical Sciences, adopted in 1982 and amended in 1993 and 2002

Recognizing that this Declaration is to be understood in a manner consistent with domestic and international law in conformity with human rights law

Recalling the Constitution of UNESCO adopted on 16 November 1945

Considering UNESCO's role in identifying universal principles based on shared ethical values to guide scientific and technological development and social transformation in order to identify emerging challenges in science and technology taking into account the responsibility of the present generations towards future generations, and that questions of bioethics, which necessarily have an international dimension, should be treated as a whole, drawing on the principles already stated in the Universal Declaration on the Human Genome and Human Rights and the International Declaration on Human Genetic Data and taking account not only of the current scientific context but also of future developments

Aware that human beings are an integral part of the biosphere, with an important role in protecting one another and other forms of life, in particular animals

Recognizing that, based on the freedom of science and research, scientific and technological developments have been, and can be, of great benefit to humankind in increasing, inter alia, life expectancy and improving the quality of life, and emphasizing that such developments should always seek to promote the welfare of individuals, families, groups or communities and humankind as a whole in the recognition of the dignity of the human person and universal respect for, and observance of, human rights and fundamental freedoms

Recognizing that health does not depend solely on scientific and technological research developments but also on psychosocial and cultural factors

Also recognizing that decisions regarding ethical issues in medicine, life sciences and associated technologies may have an impact on individuals, families, groups or communities and humankind as a whole

Bearing in mind that cultural diversity, as a source of exchange, innovation and creativity, is necessary to humankind and, in this sense, is the common heritage of humanity, but emphasizing that it may not be invoked at the expense of human rights and fundamental freedoms

Also bearing in mind that a person's identity includes biological, psychological, social, cultural and spiritual dimensions

Recognizing that unethical scientific and technological conduct has had a particular impact on indigenous and local communities

Convinced that moral sensitivity and ethical reflection should be an integral part of the process of scientific and technological developments and that bioethics should play a predominant role in the choices that need to be made concerning issues arising from such developments

Considering the desirability of developing new approaches to social responsibility to ensure that progress in science and technology contributes to justice, equity and to the interest of humanity

Recognizing that an important way to evaluate social realities and achieve equity is to pay attention to the position of women

Stressing the need to reinforce international cooperation in the field of bioethics, taking into account, in particular, the special needs of developing countries, indigenous communities and vulnerable populations

Considering that all human beings, without distinction, should benefit from the same high ethical standards in medicine and life science research

Proclaims the principles that follow and **adopts** the present Declaration.

General provisions

Article 1 – Scope

1 This Declaration addresses ethical issues related to medicine, life sciences and associated technologies as applied to human beings, taking into account their social, legal and environmental dimensions.
2 This Declaration is addressed to States. As appropriate and relevant, it also provides guidance to decisions or practices of individuals, groups, communities, institutions and corporations, public and private.

Article 2 – Aims

The aims of this Declaration are:

(a) to provide a universal framework of principles and procedures to guide States in the formulation of their legislation, policies or other instruments in the field of bioethics;
(b) to guide the actions of individuals, groups, communities, institutions and corporations, public and private;
(c) to promote respect for human dignity and protect human rights, by ensuring respect for the life of human beings, and fundamental freedoms, consistent with international human rights law;
(d) to recognize the importance of freedom of scientific research and the benefits derived from scientific and technological developments, while stressing the need for such research and developments to occur within the framework of ethical principles set out in this Declaration and to respect human dignity, human rights and fundamental freedoms;
(e) to foster multidisciplinary and pluralistic dialogue about bioethical issues between all stakeholders and within society as a whole;
(f) to promote equitable access to medical, scientific and technological developments as well as the greatest possible flow and the rapid sharing of knowledge concerning those developments and the sharing of benefits, with particular attention to the needs of developing countries;
(g) to safeguard and promote the interests of the present and future generations;
(h) to underline the importance of biodiversity and its conservation as a common concern of humankind.

Principles

Within the scope of this Declaration, in decisions or practices taken or carried out by those to whom it is addressed, the following principles are to be respected.

Article 3 – Human dignity and human rights

1. Human dignity, human rights and fundamental freedoms are to be fully respected.
2. The interests and welfare of the individual should have priority over the sole interest of science or society.

Article 4 – Benefit and harm

In applying and advancing scientific knowledge, medical practice and associated technologies, direct and indirect benefits to patients, research participants and other affected individuals should be maximized and any possible harm to such individuals should be minimized.

Article 5 – Autonomy and individual responsibility

The autonomy of persons to make decisions, while taking responsibility for those decisions and respecting the autonomy of others, is to be respected. For persons who are not capable of exercising autonomy, special measures are to be taken to protect their rights and interests.

Article 6 – Consent

1. Any preventive, diagnostic and therapeutic medical intervention is only to be carried out with the prior, free and informed consent of the person concerned, based on adequate information. The consent should, where appropriate, be express and may be withdrawn by the person concerned at any time and for any reason without disadvantage or prejudice.
2. Scientific research should only be carried out with the prior, free, express and informed consent of the person concerned. The information should be adequate, provided in a comprehensible form and should include modalities for withdrawal of consent. Consent may be withdrawn by the person concerned at any time and for any reason without any disadvantage or prejudice. Exceptions to this principle should be made only in accordance with ethical and legal standards adopted by States, consistent with the principles and provisions set out in this Declaration, in particular in Article 27, and international human rights law.
3. In appropriate cases of research carried out on a group of persons or a community, additional agreement of the legal representatives of the group or community concerned may be sought. In no case should a collective community agreement or the consent of a community leader or other authority substitute for an individual's informed consent.

Article 7 – Persons without the capacity to consent

In accordance with domestic law, special protection is to be given to persons who do not have the capacity to consent:

(a) authorization for research and medical practice should be obtained in accordance with the best interest of the person concerned and in accordance with domestic law. However, the person concerned should be involved to the greatest extent possible in the decision-making process of consent, as well as that of withdrawing consent;

(b) research should only be carried out for his or her direct health benefit, subject to the authorization and the protective conditions prescribed by law, and if there is no research alternative of comparable effectiveness with research participants able to consent. Research which does not have potential direct health benefit should only be undertaken by way of exception, with the utmost restraint, exposing the person only to a minimal risk and minimal burden and, if the research is expected to contribute to the health benefit of other persons in the same category, subject to the conditions prescribed by law and compatible with the protection of the individual's human rights. Refusal of such persons to take part in research should be respected.

Article 8 – Respect for human vulnerability and personal integrity

In applying and advancing scientific knowledge, medical practice and associated technologies, human vulnerability should be taken into account. Individuals and groups of special vulnerability should be protected and the personal integrity of such individuals respected.

Article 9 – Privacy and confidentiality

The privacy of the persons concerned and the confidentiality of their personal information should be respected. To the greatest extent possible, such information should not be used or disclosed for purposes other than those for which it was collected or consented to, consistent with international law, in particular international human rights law.

Article 10 – Equality, justice and equity

The fundamental equality of all human beings in dignity and rights is to be respected so that they are treated justly and equitably.

Article 11 – Non-discrimination and non-stigmatization

No individual or group should be discriminated against or stigmatized on any grounds, in violation of human dignity, human rights and fundamental freedoms.

Article 12 – Respect for cultural diversity and pluralism

The importance of cultural diversity and pluralism should be given due regard. However, such considerations are not to be invoked to infringe upon human dignity, human rights and fundamental freedoms, nor upon the principles set out in this Declaration, nor to limit their scope.

Article 13 – Solidarity and cooperation

Solidarity among human beings and international cooperation towards that end are to be encouraged.

Article 14 – Social responsibility and health

1. The promotion of health and social development for their people is a central purpose of governments that all sectors of society share.
2. Taking into account that the enjoyment of the highest attainable standard of health is one of the fundamental rights of every human being without distinction of race, religion, political belief, economic or social condition, progress in science and technology should advance:

(a) access to quality health care and essential medicines, especially for the health of women and children, because health is essential to life itself and must be considered to be a social and human good;
(b) access to adequate nutrition and water;
(c) improvement of living conditions and the environment;
(d) elimination of the marginalization and the exclusion of persons on the basis of any grounds;
(e) reduction of poverty and illiteracy.

Article 15 – Sharing of benefits

1 Benefits resulting from any scientific research and its applications should be shared with society as a whole and within the international community, in particular with developing countries. In giving effect to this principle, benefits may take any of the following forms:

(a) special and sustainable assistance to, and acknowledgement of, the persons and groups that have taken part in the research;
(b) access to quality health care;
(c) provision of new diagnostic and therapeutic modalities or products stemming from research;
(d) support for health services;
(e) access to scientific and technological knowledge;
(f) capacity-building facilities for research purposes;
(g) other forms of benefit consistent with the principles set out in this

Declaration.

2 Benefits should not constitute improper inducements to participate in research.

Article 16 – Protecting future generations

The impact of life sciences on future generations, including on their genetic constitution, should be given due regard.

Article 17 – Protection of the environment, the biosphere and biodiversity

Due regard is to be given to the interconnection between human beings and other forms of life, to the importance of appropriate access and utilization of biological and genetic resources, to respect for traditional knowledge and to the role of human beings in the protection of the environment, the biosphere and biodiversity.

Application of the principles

Article 18 – Decision-making and addressing bioethical issues

1 Professionalism, honesty, integrity and transparency in decision-making should be promoted, in particular declarations of all conflicts of interest and appropriate sharing of knowledge. Every endeavour should be made to use the best available scientific knowledge and methodology in addressing and periodically reviewing bioethical issues.
2 Persons and professionals concerned and society as a whole should be engaged in dialogue on a regular basis.
3 Opportunities for informed pluralistic public debate, seeking the expression of all relevant opinions, should be promoted.

Article 19 – Ethics committees

Independent, multidisciplinary and pluralist ethics committees should be established, promoted and supported at the appropriate level in order to:

(a) assess the relevant ethical, legal, scientific and social issues related to research projects involving human beings;
(b) provide advice on ethical problems in clinical settings;
(c) assess scientific and technological developments, formulate recommendations and contribute to the preparation of guidelines on issues within the scope of this Declaration;
(d) foster debate, education and public awareness of, and engagement in, bioethics.

Article 20 – Risk assessment and management

Appropriate assessment and adequate management of risk related to medicine, life sciences and associated technologies should be promoted.

Article 21 – Transnational practices

1 States, public and private institutions, and professionals associated with transnational activities should endeavour to ensure that any activity within the scope of this Declaration, undertaken, funded or otherwise pursued in whole or in part in different States, is consistent with the principles set out in this Declaration.
2 When research is undertaken or otherwise pursued in one or more States (the host State(s)) and funded by a source in another State, such research should be the object of an appropriate level of ethical review in the host State(s) and the State in which the funder is located. This review should be based on ethical and legal standards that are consistent with the principles set out in this Declaration.
3 Transnational health research should be responsive to the needs of host countries, and the importance of research contributing to the alleviation of urgent global health problems should be recognized.
4 When negotiating a research agreement, terms for collaboration and agreement on the benefits of research should be established with equal participation by those party to the negotiation.
5 States should take appropriate measures, both at the national and international levels, to combat bioterrorism and illicit traffic in organs, tissues, samples, genetic resources and genetic-related materials.

Promotion of the Declaration

Article 22 – Role of States

1 States should take all appropriate measures, whether of a legislative, administrative or other character, to give effect to the principles set out in this Declaration in accordance with international human rights law. Such measures should be supported by action in the spheres of education, training and public information.
2 States should encourage the establishment of independent, multidisciplinary and pluralist ethics committees, as set out in Article 19.

Article 23 – Bioethics education, training and information

1 In order to promote the principles set out in this Declaration and to achieve a better understanding of the ethical implications of scientific and technological developments, in particular for young people, States should endeavour

to foster bioethics education and training at all levels as well as to encourage information and knowledge dissemination programmes about bioethics.

2 States should encourage the participation of international and regional intergovernmental organizations and international, regional and national non governmental organizations in this endeavour.

Article 24 – International cooperation

1 States should foster international dissemination of scientific information and encourage the free flow and sharing of scientific and technological knowledge.

2 Within the framework of international cooperation, States should promote cultural and scientific cooperation and enter into bilateral and multilateral agreements enabling developing countries to build up their capacity to participate in generating and sharing scientific knowledge, the related know-how and the benefits thereof.

3 States should respect and promote solidarity between and among States, as well as individuals, families, groups and communities, with special regard for those rendered vulnerable by disease or disability or other personal, societal or environmental conditions and those with the most limited resources.

Article 25 – Follow-up action by UNESCO

1 UNESCO shall promote and disseminate the principles set out in this Declaration. In doing so, UNESCO should seek the help and assistance of the Intergovernmental Bioethics Committee (IGBC) and the International Bioethics Committee (IBC).

2 UNESCO shall reaffirm its commitment to dealing with bioethics and to promoting collaboration between IGBC and IBC.

Final provisions

Article 26 – Interrelation and complementarity of the principles

This Declaration is to be understood as a whole and the principles are to be understood as complementary and interrelated. Each principle is to be considered in the context of the other principles, as appropriate and relevant in the circumstances.

Article 27 – Limitations on the application of the principles

If the application of the principles of this Declaration is to be limited, it should be by law, including laws in the interests of public safety, for the investigation, detection and prosecution of criminal offences, for the protection of public health or for the protection of the rights and freedoms of others. Any such law needs to be consistent with international human rights law.

Article 28 – Denial of acts contrary to human rights, fundamental freedoms and human dignity

Nothing in this Declaration may be interpreted as implying for any State, group or person any claim to engage in any activity or to perform any act contrary to human rights, fundamental freedoms and human dignity.

Bibliography

Abbott F., Cottier T., Gurry F., *The International Intellectual Property System: Commentary and Materials, Part I*, The Hague: Kluwer Law International, 2003.

Alexy R., 'Effects of Defects – Action or Argument? Thoughts about Deryck Beyleveld and Roger Brownsword's Law as a Moral Judgment', *Ratio Juris* 19 (2), 2006, 169–79.

Aljalian N. N., 'The Role Of Patent Scope In Biopharmaceutical Patents', *Boston University Journal of Science and Technology Law* 11, 2005, 1–76.

Alston P., 'Ships Passing in the Night: The Current State of the Human Rights and Development Debate Seen Through the Lens of the Millennium Development Goals', *Human Rights Quarterly* 27 (3), 2005, 755–829.

Alston P. (ed.), *Peoples' Rights,* Oxford: Oxford University Press, 2001.

Alston P., Tomasevski K. (eds), *The Right to Food,* Netherlands: SIM, 1984.

Amin S., *After the New International Economic Order: The Future of International Economic Relations*, UNCTAD Trade and Development Review, 1982, 9–23.

Anonymous, 'Human Rights in the Horn Of Africa', *Indiana International and Comparative Law Review* 11, 2001, 543–61.

Annas G. J., 'Mengele's Birthmark: The Nuremberg Code in United States Courts', *Journal of Contemporary Health Law & Policy* 7, 1991, 17–45.

Annas G. J., 'American Bioethics and Human Rights: The End of all Our Exploring', *Journal of Law, Medicine & Ethics* 32, 2004, 658–62.

Annas G., Glantz L. H., Katz B. F. (eds) *Informed Consent to Human Experimentation: The Subject's Dilemma*, Cambridge, Massachusetts: Ballinger, 1977.

Araujo R. J., 'The Meaning of Person in the Context of Human Embryonic Cloning – Evolving Challenges for the Rule of Law in the International Order', *University of Saint Thomas Law Journal* 1, 2003, 39–55.

Ari Patrinos A., Drell D. W., 'Introducing the Human Genome Project: Its Relevance, Triumphs, and Challenges', *Judges' Journal* 36 (3), 1997, 5–10.

Arsanjani M., 'Negotiating the UNDeclaration on Human Cloning', *The American Journal of International Law* 100 (1), 2006, 164–179.

Asai A., Oe S., 'A Valuable Up-to-Date Compendium of Bioethical Knowledge', *Developing World Bioethics* 5 (3), 2005, 218.

Barsh R. L., 'The right to development as a human right: Results of the Global Consultation', *Human Rights Quarterly* 13, 1991, 322–38.

Baslar K., *The Concept of Common Heritage of Mankind in International Law*, The Hague/Boston/London: Martinus Nijhoff Publishers, 1998.

Baxi U., 'Voices Of Suffering And The Future Of Human Rights', *Transnational Law and Contemporary Problems* 8, 1998, 125–69.

Beauchamp T. L., 'Methods and Principles in Biomedical Ethics', *Journal of Medical Ethics* 29, 2003, 269–74.

Beauchamp T. L., Childress J. F., *Principles of Biomedical Ethics*, New York: Oxford University Press, 1st edn, 1979 (5th edn, 2001).

Bedjaoui M., *Towards a New International Economic Order*, Paris: UNESCO, 1979.

Behrman J. N., *Toward a New International Economic Order*, Paris: The Atlantic Institute for International Affairs, 1974, 68–9.

Beyleveld D., *The Dialectical Necessity of Morality: An Analysis and Defense of Alan Gewirth's Argument to the Principle of Generic Consistency*, Chicago: Chicago University Press, 1991.

Beyleveld D, Brownsword R., 'Human Dignity, Human Rights, and Human Genetics', *The Modern Law Review* 61, 1998, 661–81.

—— *Human Dignity in Bioethics and Biolaw*, Oxford: Oxford University Press, 2001.

Bertelli A., 'Impoverished Liberalism: Does the New York Workfare Program Violate Human Rights?', *Buffalo Human Rights Law Review* 5, 1999, 175–209.

Bhabha J., 'The Right To Community?', *University of Chicago Law Review* 64, 1997, 1117–28.

Bhagwati J.N. (ed.), *The New International Economic Order: The North-South Debate*, Cambridge, Massachusetts/London: The MIT Press, 1977.

Birnbaum L., 'Medical Ethics in Exercise Physiology', *Journal of Professional Exercise Physiology* 10 (3), published online October 2005 and available at http://www.exercisephysiologists.com/MedicalEthics/index.html.

Blackburn R., Taylor J. (eds), *Human Rights for the 1990s*, London: Mansell Publishing Limited, 1991 (repr. 1991, 1992).

Bohrer R. A., 'Foreword: What is Biotechnology and Why Devote a Law Review Symposium to Biotechnology Law?', *University of Pittsburgh Law Review* 55, 1994, 607–10.

Bond J. E., 'International Intersectionality: A Theoretical and Pragmatic Exploration of Women's International Human Rights Violations', *Emory Law Journal* 52, 2003, 71–152.

Bovenberg J. A., 'Mining the Common Heritage of Our DNA: Lessons Learned from Grotius And Pardo', *Duke Law & Technology Review* 0008, 2006.

Brandt W., *North South: A Programme for Survival*, USA: The MIT Press, 1980.

Brown E. D., *Sea-bed Energy and Minerals: The International Legal Regime, Vol.2: Sea-Bed Mining*, The Hague: Kluwer Law International, 2001.

Cassese A., *International Law*, 2nd edn, Oxford: Oxford University Press, 2005.

Caulfield T., 'Human cloning laws, human dignity and the poverty of the policy making dialogue', *BMC Medical Ethics* 4 (3), 2003.

Cavalli-Sforza L., 'The Human Genome Diversity Project: past, present and future', *Nature Reviews Genetics* 6 (4), 2005, 333–40.

Charlesworth H., *Writings in Rights: Australia and the Protection of Human Rights*, Sydney: University of New South Wales Press Limited, 2002.

Charney J. I., 'Entry into Force of the 1982 Convention on the Law of the Sea', *Virginia Journal of International Law* 35, 1995, 381–404.

Chauhan S., *GATT to WTO: Gandhian Alternative to New International Economic Order*, New Delhi: Deep and Deep Publications Pvt Ltd, 2001

Chesire W. P., 'Human Embryo Research and the Language of Moral Uncertainty', *The American Journal of Bioethics* 4 (1), 2004, 1–5.

Chinkin C. M., 'Human Rights in Global Politics', *The American Journal of International Law*, 95 (2), 2001, 472–7.

Chowdhury S. R., Denters E. M. J., de Waart P. J. I. M., *The Right to Development in International Law*, Dordrecht: Martinus Nijhoff Publishers, 1992.

Commission on Health Research for Development, *Health Research: Essential Link to Equity in Development*, New York: Oxford University Press, 1990.

Cook-Deegan R., *The Gene Wars: Science, Politics and the Human Genome*, New York: W. W. Norton and Company, 1994.

—— 'Origins Of The Human Genome Project', *Risk: Health, Safety & Environment*, 1994, 97–118.

Copelon R., 'The Indivisible Framework of International Human Rights: A Source of Social Justice in the U.S.', *New York City Law Review*, May 3, 1998, 59–80.

Corea G., *Need for Change: Towards the New International Economic Order*, Oxford: Pergamon Press, 1980.

Cornish and Llewelyn, *Intellectual Property: Patents, Copyright, Trademarks and Allied Rights*, London: Thompson, Sweet and Maxwell, 5th edn, 2003.

Craven M. C. R., *The International Covenant on Economic, Social, and Cultural Rights: A Perspective on its Development*, Oxford: Clarendon Press, 1995.

Daniels N., 'Technology and Resource Allocation: Old Problems in New Clothes', *Southern California Law Review* 65, 1991, 225–40.

Dias C., *Development, Human Rights and the Rule of Law*, New York: Pergamon Press, 1981.

Dinwoodie G. B., 'The International Intellectual Property Law System: New Actors, New Institutions, New Sources', *Marquette Intellectual Property Law Review*, Special Issue 10, 2006, 205.

Division of Ocean Affairs and the Law of the Sea, Office of Legal Affairs, *The Law of the Sea: Concept of the Common Heritage of Mankind- Legislative History of Article 133 to 150 and 311(6) of the United Nations Convention on the Law of the Sea*, New York: United Nations, 1996

Dupuy R., 'The Right to Development at the International Level', Workshop, The Hague, 16–18 October 1979, The Hague Academy of International Law, Alphen aan den Rijn, The Netherlands: Sijthoff and Noordhoff, 1980.

Dworkin R., *Taking Rights Seriously*, Boston: Harvard University Press, 1977.

Dyer O., 'US awards patent for tribesman's DNA', *British Medical Journal*, 311, 1995, 1452.

Edelman P. B., 'The Next Century of Our Constitution: Rethinking Our Duty to the Poor', *Hastings Law Journal* 39, 1987, 1–61.

Editorial: Proteomics' new order, *Nature* 437, 2005, 169–70.

Elshtain J. B., 'Law and Moral Life', *Yale Journal of Law and Humanities* 11, 1999, 383–413.

Engle E. A., 'The Transformation of the International Legal System: The Post-Westphalian Legal Order', *Queensland Law Review* 23, 2004, 23–45.

Erasmus M. G., *The New International Economic Order and International Organisations: Towards a Special Status for Developing Countries?*, Frankfurt/Main: Haag und Herchen, 1979.

Faunce T. A., 'Will international human rights subsume medical ethics? Intersections in the UNESCO Universal Bioethics Declaration', *Journal of Medical Ethics* 31, 2005, 173–8.

Ferrero R., Special Rapporteur of the Sub-Commission on Prevention of Discrimination and Protection of Minorities, *The New International Economic Order and the Promotion of Human Rights*, New York: United Nations, 1986.

Fiandaca S., 'In Vitro Fertilization and Embryos: The Need for International Guidelines', *Albany Law Journal of Science & Technology* 8, 1998, 337–404.

Fidler D. P., *International Law and Infectious Diseases*, Oxford: Clarendon Press, 1999.

Fitzgerald B. F., 'Trade-based Constitutionalisms: The Framework for Universalizing Substantive International Law?', *University of Miami Yearbook of International Law* 97 (5), 1996, 111–63.

Francioni F. (ed.), *Biotechnologies and International Human Rights*, Oxford: Hart Publishing, 2007.

Franck T., Munansangu M., *The New International Economic Order: International Law in the Making?*, New York: United Nations Institute for Training and Research, 1982.

Frankel M., *High Noon in the Cold War: Kennedy, Krushchev, and the Cuban Missile Crisis*, USA: Presidio Press, repr. 2005.

Freedman L., *Kennedy's Wars: Berlin, Cuba, Laos, and Vietnam*, New York: Oxford University Press, 2000.

French D., '"From Seoul with Love" – the continuing relevance of the 1986 Seoul ILA Declaration on Progressive Development of Principles of Public International Law relating to a New International Economic Order', *Netherlands International Law Review* 55, 2008, 3–31.

Gárdián G., Vécsei L., 'Huntington's disease: pathomechanism and therapeutic perspectives', *Journal of Neural Transmission* 111 (10,11), 2004, 1485–94.

George E. P., 'The Stem Cell Debate: The Legal, Political and Ethical Issues Surrounding Federal Funding of Scientific Research on Human Embryos', *Albany Law Journal of Science & Technology*, 12 (2), 2002, 747–800.

Gerhart P. M., 'Why law making for global intellectual property is unbalanced', *European Intellectual Property Review* 22 (7), 2000, 309–13.

Gevurtz F. A., 'Biotechnology and the Law', *McGeorge Law Review* 32, 2000, 85–7.

Gewirth A., *Reason and Morality*, Chicago: University of Chicago Press, 1978.

—— 'The Basis and Content of Human Rights', *Georgia Law Review* 13, 1979, 11–43.

—— *Human Rights: Essays on Justification and Application*, Chicago: University of Chicago Press, 1982.

—— *The Community of Rights*, Chicago: The University of Chicago Press, 1996.

Ghosh P. K. (ed.), *Third World Development: A Basic Needs Approach*, Connecticut: Greenwood Press, 1984.

Giarelli E., 'Safeguarding being: a bioethical principle for genetic nursing care', *Nursing Ethics* 10 (3), 2003, 255–68.

Gillon R., 'Medical ethics: four principles plus attention to scope', *British Medical Journal* 309, 1994, 184.

—— 'Ethics needs principles – four can encompass the rest – and respect for autonomy should be "first among equals"', *Journal of Medical Ethics* 29, 2003, 307–12.

Gitter D. M., 'Resolving the Open Source Paradox in Biotechnology: A Proposal for a Revised Open Source Policy for Publicly Funded Genomic Databases', *Houston Law Review* 43, 2007, 1475–1521.

Glasner P., Atkinson P., 'The Global Genome: Biotechnology, Politics and Culture', *Body and Society* 12 (3), 2006, 123–31.

Goldie L. F. E., 'A Note on Some Diverse Meanings of the Common Heritage of Mankind', *Syracuse Journal of International Law and Commerce* 10, 1983, 69–112.

Goldstein W., 'Despair and the UN Development Decade', *Futures* magazine, 1981, pp. 206–16.

Gore A., 'Planning a New Biotechnology Policy', *Harvard Journal of Law and Technology* 5, 1991, 19–30.

Gorove S., 'The Concept of Common Heritage of Mankind: A Political, Moral or Legal Innovation?', *San Diego Law Review* 9, 1972, 390–403.

Gostin O., Lazzarini Z., *Human Rights and Public Health in the AIDS Pandemic*, Oxford: Oxford University Press, 1997.

Grisolia S., 'Mapping the Human Genome', *The Hastings Center Report* 19 (4), 1989, 18–19.

Gross L. (ed.), *The American Society of International Law*, New York: Appleton-Century-Crofts, 1969.

Haas A. K., 'The Wellcome Trust's Disclosures of Gene Sequence Data into the Public Domain and the Potential for Proprietary Rights in the Human Genome', *Berkeley Technology Law Journal* 16, 2001, 145–64.

Hamilton N. D., 'Who Owns Dinner: Evolving Legal Mechanisms for Ownership of Plant Genetic Resources', *Tulsa Law Journal* 28, 1993, 587–657.

Hanauske-Abel H. M., 'Not a slippery slope or sudden subversion: German medicine and National Socialism in 1933', *British Medical Journal* 313, 1996, 1453–63.

Hanke J., 'Genomics and New Technologies as Catalysts for Change in the Drug Discovery Paradigm', *Journal of Law, Medicine and Ethics* 28, 2000, 15–22.

Hannum H., 'Human Rights as Politics and Idolatry', *The American Journal of International Law*, 97 (4), 2003, 1006–9

Hardcastle R., *Law and the Human Body: Property Rights, Ownership and Control*, Oxford: Hart Publishing.

Henkin L., Pugh R., Schachter O., Smit H., *International Law*, St. Paul, Minnesota: West Publishing Co., 2nd edn, 1987.

Holm S., 'Book Review: Principles of Biomedical Ethics', 5th edn, *Journal of Medical Ethics 28*, 2002, 332.

Huxley J., *UNESCO: Its Purpose and its Philosophy*, Washington DC: Public Affairs Press, 1947.

Hyman D. A., 'Does Technology Spell Trouble with a Capital "T"?: Human Dignity and Public Policy', *Harvard Journal of Law and Public Policy* 27, 2003, 3–18.

Isasi R. M., Knoppers B. M., Singer P. A., Daar A. S., 'Legal and Ethical Approaches to Stem Cell and Cloning Research: A Comparative Analysis of Policies in Latin America, Asia, and Africa', *Journal of Law, Medicine & Ethics* 32, 2004, 626–37.

Isfahan M., Oosterveld V. (eds), *Giving Meaning to Economic, Social and Cultural Rights*, Philadelphia: University of Philadelphia Press, 2001.

Jasanoff S. (ed.), *States of Knowledge: The Co-production of Science and Social Order*, London: Routledge, 2004.

Jenks C. W., 'Economic and Social Change and the Law of Nations', *Receuil des Cours*, 138, 1973.

Johnson H. G., *The New International Economic Order*, Graduate School of Business: Selected Papers No. 49, University of Chicago, 1976.

Jones J. H., *Bad Blood: The Tuskegee Syphilis Experiment*, Free Press, rev. edn, 1993.

Jonsen A. R., *The Birth of Bioethics*, New York: Oxford University Press, 1998.

Jonsen A. R, Veatch, R. M., Walters L. (eds), *Source Book in Bioethics: A Documentary History*, Georgetown: Georgetown University Press, 1998.

Kahn S. L., 'The Right to Adequate Treatment Versus the Right to Refuse Antipsychotic Drug Treatment: A Solution to the Dilemma of the Involuntarily Committed Psychiatric Patient', *Emory Law Journal* 33, 1984, 441–85.

Kanellopoulos G., 'Embryonic Stem Cell Research: A Comparative Study of the Philosophies of the United States and the United Kingdom', *Journal of International Business and Law* 4, 2005, 170–91.

Keller L. M., *The Indivisibility of Economic and Political Rights*, Book Review of A. Sen, *Development as Freedom*, New York: Knopf, 1999.

King Jr H. T., Nuremberg and Sovereignty, *Case Western Reserve Journal of International Law* 28, 1996, 135–40.

King R. A., Rotter J. I., Motulsky A. G., *The Genetic Bases of Common Diseases*, Oxford: Oxford University Press, 2002.

Landman W., Schüklenk U., 'UNESCO "declares" universals on bioethics and human rights – many unexpected universal truths unearthed by UN body', *Developing World Bioethics* 5 (3), 2005, iii–vi.

Leaning J., 'War crimes and medical science', *British Medical Journal* 1996: 313 (7070), 1413–15.

Lenoir N., 'Universal Declaration on the Human Genome and Human Rights: The First Legal and Ethical Framework at the Global Level', *Columbia Human Rights Law Review* 30, 1999, 537–87.

Lessig L., *The Future of Ideas: The Fate of the Commons in a Connected World*, New York: Random House, 2001.

Lillich R. B. (ed.), *Economic Coercion and New International Economic Order*, Charlottesville, Virginia: MC Law Publishers, 1976.

Looney R., 'New International Economic Order', in Jones, R. J. B. (ed.), *Encyclopedia of International Political Economy*, London: Routledge, 1999.

McCorquodale R. (ed.), *Human Rights (International Library of essays in law and legal theory. Second Series)*, Aldershot, England: Ashgate-Dartmouth, 2003.

Macer R. J. D., *Bioethics is Love of Life: An Alternative Textbook*, Christchurch, New Zealand: Eubios Ethics Institute, 1998.

McFarlane A., 'In the Business of Development: Development Policy in the First Two Years of the Bush Administration', *Berkeley Journal of International Law* 21, 2003, 521–47.

McFarland-Icke B., 'Nazi Medicine and the Nuremberg Trials', *Social History of Medicine* 18 (3), 2005, 523–4.

Macklin R., 'Yet another Guideline? The UNESCO Draft Declaration', *Developing World Bioethics*, 5 (3), 2005, 244–50.

Macqueen H., Waelde, C., Laurie, G., *Contemporary Intellectual Property: Law and Policy*, Oxford: Oxford University Press, 2007.

Marks S., *Obstacles to the Right to Development*, François-Xavier Bagnoud Center for Health and Human Rights, 2003.

Marks S., Clapham A., *International Human Rights Lexicon*, Oxford: Oxford University Press, 2005.

Mauron A., 'Is the genome the secular equivalent of the soul?', *Science* 291 (5505), 2001, 831–2.

M'Baye K., 'Le droit au développement comme UNdroit de l'homme', *Human Rights Journal* 5, 1972, 503–34.

M'Charek A., *The Human Genome Diversity Project: An Ethnography of Scientific Practice*, Cambridge: Cambridge University Press, 2005.

Meron T., 'On a Hierarchy of International Human Rights', *American Journal of International Law* 80, January 1986, 1–23.

Meslin M., 'Religious Traditions and the Human Person', in H. G. Kippenberg, Y. B. Kuiper, and A.F. Sanders (eds), *Concepts of Person in Religion and Thought*, Berlin: Mouton de Gruyter, 1990.

Moreno J. D., 'Reassessing the Influence of the Nuremberg Code on American Medical Ethics', *Journal of Contemporary Health Law & Policy* 13, 1997, 347–60.

Moses L. B., 'Understanding Legal Responses to Technological Change: The Example of In Vitro Fertilization', *Minnesota Journal of Law, Science & Technology*, 6, 2006, 505–618.
Mutua M., 'Theory and Reality in the International Protection of Human Rights', *The American Journal of International Law*, 95 (1), 2001, 255–7.
Nelson J. C., 'The Contemporary Seabed Mining Regime: A Critical Analysis of the Mining Regulations Promulgated by the International Seabed Authority', *Colorado Journal of International Environmental Law and Policy*, 16, 2005, 27–75.
Nolan-Stevaux K. M., 'Open Source Biology: A Means to Address the Access and Research Gaps?, *Santa Clara Computer and High Technology Law Journal* 23, 2007, 271–316.
Nys, H., 'Towards an International Treaty on Human Rights and Biomedicine? Some Reflections Inspired by UNESCO's Universal Declaration on Bioethics and Human Rights', *European Journal of Health Law* 13, 2005, 5–8.
Ohlin J. D., 'Is The Concept of the Person Necessary for Human Rights?', *Columbia Law Review* 105, 2005, 209–49.
Orlowski V., 'Promising Protection Through Internationally Derived Duties', *Cornell International Law Journal* 36, 2003, 381–413.
Ott A. F., 'The Impact of the New International Economic Order on the Third World' in Tyler, W. G., *Issues and Prospects for the New International Economic Order*, Toronto: Lexington Books, 1977.
Otto D., 'Rethinking the "Universality" of Human Rights Law', *Columbia Human Rights Law Review* 29, 1997, 1–46.
Pardo A., 'Law of the Sea Conference – What Went Wrong', in Freidheim R. L. (ed.), *Managing Ocean Resources: A Primer*, pp. 137–9, cited in Goldie L. F. E., 'A Note on Some Diverse Meanings of the Common Heritage of Mankind', *Syracuse Journal of International Law and Commerce* 10, 1983, 69–112 at 87.
Pennisi E., 'Finally the Book of Life and the Instructions for Navigating It', *Science* 288 (5475), 2000, 2304–7.
Petersmann E., 'On "Indivisibility" of Human Rights', *European Journal of International Law* 14 (2), 2003, 381–5.
Price D., 'The Human Tissue Act 2004', *The Modern Law Review* 68 (5), 2005.
Pridan-Frank S., 'Human-Genomics: A Challenge to the Rules of the Game of International Law', *Columbia Journal of Transnational Law* 40, 2002, 619–76.
Post S. (ed.), *Encyclopedia of Bioethics*, 3rd edn, Macmillan Reference, 2004.
Pyarelal, *Mahatma Gandhi: The Last Phase*, 2 vols, Ahmedabad, India: Navajivan Press, 1958.
Rajagopal R., 'The Allure of Normativity', *Harvard Human Rights Journal* 11, 1998, 363–77.
Reaume D. G., 'Discrimination and Dignity', *Louisiana Law Review* 63, 2003, 645–95.
Reich W. T. (ed.), *The Encyclopedia of Bioethics*, rev. edn, New York: Simon Schuster Macmillan, 1995.
Ridder S., Woll L., 'Transforming the Grounds: Autonomy and Reproductive Freedom', *Yale Journal of Law and Feminism* 2, 1989, 75–98.
Rifkin J., *The Biotech Century: Harnessing the Gene and Remaking the World*, London: Victor Gollancz, 1998.
Rittberger V. (ed.), *Global Governance and the United Nations System*, Tokyo: United Nations University Press, 2002.
Rose G., 'International Law of Sustainable Agriculture in the 21st Century: The International Treaty on Plant Genetic Resources for Food and Agriculture', *Georgetown International Environmental Law Review* 15, 2003, 583–632.

Ruger J. P., 'Toward a Theory of a Right to Health: Capability and Incompletely Theorized Agreements', *Yale Journal of Law and the Humanities*, 18, 2006, 273–326.

Salbu S. R., 'Aids and Drug Pricing: In Search of a Policy', *Washington University Law Quarterly* 71, 1993, 691–734.

Salomon M. E., 'Towards a Just Institutional Order: A Commentary on the First Session of the UN Task Force on the Right to Development', *Netherlands Quarterly of Human Rights* 23, 2005, 409–38.

Sands P., Klein P., *Bowett's Law of International Institutions*, 5th edn, London: Sweet and Maxwell, 2001.

Schachter O., 'The Evolving International Law of Development', *Columbia Journal of Transnational Law* 15, 1976, 1–16

—— 'Human Dignity as a Normative Concept', *The American Journal of International Law* 77, 1983, 848–54.

Schrödinger E., *What is Life?: The Physical Aspect of the Living Cell*, Cambridge: Cambridge University Press, 1944.

Schroeder D., 'Human Rights and their Role in Global Bioethics', *Cambridge Quarterly of Healthcare Ethics* 14 (2), 2005, 221–3.

Schroeder D., Ladikas M., Schuklenk U., Lasen-Diaz C., Kleinsmidt A., Alvarez-Castillo F., Feinholz D., 'Sharing the benefits of genetic research', *British Medical Journal* 331, 2005, 1351–2.

Scott J. B. (ed.), Magoffin R. v D. (trans.), *Hugo Grotius, The Freedom of the Seas*, Oxford: Oxford University Press, 1916.

Scriver C. R., 'Presidential Address', *American Journal of Human Genetics* 401, 1987, 99–211.

Sen A., *Development as Freedom*, Oxford: Oxford University Press, 1999.

Sengupta A., 'On the Theory and Practice of the Right to Development', *Human Rights Quarterly* 24 (4), 2002, 837–89.

Shestack J. J., 'The Philosophic Foundations of Human Rights', *Human Rights Quarterly* 20 (2), 1998, 201–34.

Shibuya K., Ciecierski C., Guindon E., Bettcher D. W., Evans, D. B., Murray, C. J. L., 'WHO Framework Convention on Tobacco Control: development of an evidence based global public health treaty', *British Medical Journal* 327, 2003, 154–7.

Shreeve J., *The Genome War: How Craig Venter Tried to Capture the Code of Life and Save the World*, London: Knopf (Random House), 2004.

Shue H., *Basic Rights: Subsistence, Affluence, and U.S. Foreign Policy*, Princeton: Princeton University Press, 2nd edn, 1996.

Singer J. W., 'The Legal Rights Debate in Analytical Jurisprudence from Bentham to Hohfeld', *Wisconsin Law Review*1982, 975–1059.

Stanovnik J., *Towards the New International Economic Order*, Belgrade: Jugoslovenska Stavarnost, 1979.

Steiner H. J., Alston P., *International Human Rights in Context: Law, Politics and Morals*, Texts and Materials, 2nd edn, Oxford: Oxford University Press, 2000.

Subramanian N., Chung L., 'Measuring the Evolvability of Legal Personality', *Ius Gentium* 11, 2005, 79–130.

Sukijasovic M., 'The Contribution of the Charter on Economic Rights and Duties of States to the Elaboration of the Notion of "New International Economic Order"', in *Legal Aspects of a New International Economic Order*, Belgrade: Institute of Comparative Law, 1979.

Sunder M., 'Piercing the Veil', *Yale Law Journal* 112, 2003, 1399–1472.

Terrell T. P., 'Flatlaw: An Essay on the Dimensions of Legal Reasoning and the Development of Fundamental Normative Principles', *California Law Review* 72, 1984, 288–342.

Tilford D. S., 'Saving the Blueprints: The International Legal Regime for Plant Resources', *Case Western Reserve Journal of International Law* 30, 1998, 373–446.

Toebes C. A. B., *The Right to Health as a Human Right in International Law*, Antwerpen, Groningen, Oxford: Intersentia, Hart Publishers, 1999.

Tomushat C., *Human Rights: Between Idealism and Realism*, Oxford: Oxford University Press, 2003

Townend D., 'Who owns genetic information?' in Sandor, J. (ed.), *Society and Genetic Information: Codes and Laws in the Genetic Era*, Budapest: CPS-Central European University Press, 2003, p. 132.

Ul-Haq M., *Reflections on Human Development*, Oxford: Oxford University Press, India, 2nd edn, 1999.

Uvin P., *Human Rights and Development*, Connecticut: Kumarian Press Inc., 2004.

Varges G. S., *The New International Economic Order Legal Debate: Background, Status, and Alternatives*, European University Studies, vol. 326, Frankfurt am Main: Peter Lang, 1983.

Walters G. J., *Human Rights in an Information Age: A Philosophical Analysis*, Toronto: University of Toronto Press, 2001.

Warner D., 'An Ethics of Human Rights: Two Interrelated Misunderstandings', *Denver Journal of International Law and Policy* 24, 1996, 395–415.

Watson J. D., *The Molecular Biology of the Gene*, 1st edn, New York: W. A. Benjamin, 1965.

—— *The Double Helix*, London: Weidenfeld & Nicolson, 1968 (revised with a new introduction, 1997).

Watson J. D., Crick F. H. C., 'A Structure for Deoxyribose Nucleic Acid', *Nature*, 1953, 171–7.

Weindling P. J., 'Human guinea pigs and the ethics of experimentation: the *BMJ*'s correspondent at the Nuremberg medical trial', *British Medical Journal* 313 (7070), 1996, 1467–70.

—— *Nazi Medicine and the Nuremberg Trials: From Medical War Crimes to Informed Consent*, London: Palgrave Macmillan, 2004.

Weiss E. B. (ed.), *Environmental Change and International Law: New Challenges and Dimensions*, The United Nations University, Toyko: United Nations University Press, 1992.

Welsome E., *The Plutonium Files: America's Secret Medical Experiments in the Cold War*, The Dial Press, 1999.

Wilde R., 'Quis Custodiet Ipsos Custodes?: Why and How UNHCR Governance of "Development" Refugee Camps Should be Subject to International Human Rights Law', *Yale Human Rights and Development Law Journal* 1, 1998, 107–28.

Williams J. R., 'UNESCO's Proposed Declaration on Bioethics and Human Rights – A Bland Compromise', *Developing World Bioethics*, 5 (3), 2005, 210–15.

Wilner G. M., 'Reflections on Regional Human Rights Law', *Georgia Journal of International and Comparative Law*, 25, 1995/1996, 407–26.

Wolff M., 'Stop, Thief!', *New York Magazine*, 3 March 3 2003, pp. 24–6.

Wright Q., 'The Law of the Nuremberg Trial', *The American Journal of International Law*, 41 (1), 1947, 38–72.

Wuerffel K. N., 'Discrimination among Rights? A Nation's Legislating a Hierarchy of Human Rights In The Context Of International Human Rights Customary Law', *Valparaiso University Law Review* 1998, 369–412.

Wynne B., 'Public Engagement as a Means of Restoring Public Trust in Science – Hitting the Notes, but Missing the Music?', *Community Genetics* 9, 2006, 211–20.
Yamey G., 'The world's most neglected diseases – Ignored by the pharmaceutical industry and by public-private partnerships', *British Medical Journal* 325, 2002, 176–7.
Zampas C. C., Gher J. M., 'Abortion as a human right: international and regional standards', *Human Rights Law Review* 8(2), 2008, 249–94.

Treaties

Paris Convention for the Protection of Industrial Property (1883)
Treaty on Principles Governing the Activities of States in the Exploration and Use of Outer Space, Including the Moon and Other Celestial Bodies (1967)
Universal Copyright Convention (1971)
Convention on the Grant of European Patents (1973)
Agreement Governing the Activities of States on the Moon and other Celestial Bodies (1979)
The United Nations Convention on the Law of the Sea (1982)
The Convention on Biological Diversity (1992)
Marrakesh Agreement Establishing the World Trade Organization (1994)
Agreement relating to the Implementation of Part XI of the United Nations Convention on the Law of the Sea of 10 December 1982 (1994)
Mining Code (2000)
The International Treaty on Plant Genetic Resources for Food and Agriculture (2001)
WHO Framework Convention on Tobacco Control (2003)

Documents of international organisations

Commission on Human Rights Doc. RES/37/4 (1979)
Commission on Human Rights Doc. RES/33/4 (1980)
Commission on Human Rights Doc. RES/37/36 (1981)
Commission on Human Rights Doc. RES/67/91 (1993)
Commission on Human Rights Doc. RES/07 (1995)
Commission on Human Rights Doc. RES/82 (1995)
Commission on Human Rights Doc. RES/72 (1998)
Commission on Human Rights Doc. RES/63 (1999)
Commission on Human Rights Doc. RES/59/79 (1999)
Commission on Human Rights Doc. HR/CN/99/62 (1999)
Commission on Human Rights Doc. RES/69 (2002)
ECOSOC Doc. RES/40/1112 (1966)
ECOSOC Doc. E/C.12/1997/SR.27 (1997)
ECOSOC Doc. E/C.12/1998/SR.49 (1998)
ECOSOC Doc. E/CN.4/Sub.2/2004/38 (2004)
ECOSOC Doc. E/CN.4/Sub.2/2005/38 (2005)
EU Doc. 90/395/EEC (1991)
FAO Doc. RES/22/8 (1983)
FAO Doc. RES/25/4 (1989)
FAO Doc. RES/25/5 (1989)
FAO Doc. RES/26/3 (1991)

FAO Doc. RES/31/3 (2001)
UN Doc. A/RES/16/1710 (1961)
UN Doc. A/RES/16/1715 (1961)
UN Doc. A/RES/16/1710 (1961)
UN Doc. A/C.1/PV.1516 (1967).
UN Doc. A/RES/22/6695 (1967)
UN Doc. A/RES/25/2749 (1970)
UN Doc. A/RES/25/2626 (1970)
UN Doc A/RES/25/2687 (1970)
UN Doc. A/RES/27/8721 (1972)
UN Doc. A/RES/S-6/3201 (1974)
UN Doc. A/RES/S-6/3202 (1974)
UN Doc. A/RES/29/3281 (1974)
UN Doc. A/RES/30/3384 (1975)
UN Doc. A/RES/S-7/3362 (1975)
UN Doc. A/RES/34/46 (1979)
UN Doc. A/RES/34/58 (1979)
UN Doc. E/CN.4/1334 (1979)
UN Doc. A/RES/35/174 (1980)
UN Doc. ST/HR/SER.A/8 (1980)
UN Doc. A/RES/36/133 (1981)
UN Doc. A/RES/41/128 (1986)
UN Doc. A/RES/44/25, Annex, (1989)
UN Doc. A/CONF.157/23 (1993)
UN Doc. A/CONF.157/23 (1993)
UN Doc. A/CONF.157/23 (1993)
UN Doc. A/RES/48/140 (1994)
UN Doc. E/CN.4/1995/74 (1994)
UN Doc. ISBA/A/L.8 (1996) and Corr.1 (2 April 1996)
UN Doc. A/Res/51/122 (1996)
UN Doc. A/RES/53/152 (1998)
UN Doc. A/53/PV.85 (1998)
UN Doc. A/RES/53/152 (1999)
UN Doc. A/RES/54/1 (1999)
UN Doc. A/RES/55/2 (2000)
UN Doc. ISBA/6/A/18 (2000)
UN Doc. A/RES/55/2 (2000)
UN Doc. SHS-503/01/CIB-8/2 Rev. (2002)
UN Doc. E/CN.4/Sub.2/36 (2003)
UN Doc. A/RES/60/251 (2006)
UN Doc. E/CN.4/2006/26 (2006)
UN Doc. A/HRC/Sub.1/58/L.30 (2006)
UN Doc. A/RES/61/295 (2007)
UNESCO Doc. C/RES/22/13.1 (1983)
UNESCO Doc. C/RES/23/13.1 (1985)
UNESCO Doc. C/RES/23/13.1 (1985)
UNESCO Doc. C/RES/24/13.1 (1987)
UNESCO Doc. C/RES/25/5.2 and 7.3 (1989)
UNESCO Doc. C/RES/27/5.15 (1991)

UNESCO Doc. C/RES/27/45 (1993)
UNESCO Doc. C/RES/27/5.15 (1993)
UNESCO Doc. C/RES/28/0.12, 2.1 and 2.2 (1993)
UNESCO Doc. C/RES/28/0.12 (1995)
UNESCO Doc. BIO/CIB-COMJUR/95 (1995)
UNESCO Doc. C/RES/28/2.2 (1995)
UNESCO Doc. C/RES/28/ 0.12, 2.1 and 2.2 (1995)
UNESCO Doc. BIO- 97/CONF.201/3, Annex 2 (1996)
UNESCO Doc. C/RES/29/82 (1997)
UNESCO Doc. C/RES/29/16 (1997)
UNESCO Doc. BIO-7/00/GT-2/3 (Rev. 1) (2001)
UNESCO Doc. SHS-503/01/CIB-8/2 Rev. (2002)
UNESCO Doc. C/RES/32/22 (2003)
UNESCO Doc. C/RES/33/36 (2005)
WHO Doc. WPR/RC30.R13 (1979)
WHO Doc. RES/56.27/56 (2003)
WHO Doc. WHA57.13 (2004)
WHO Doc. RES/61.21/61 (2008)
WIPO Doc. WO/GA/31/11 (2004)
WIPO Doc. WO/GA/31/11, IIM/1/4 (2005)

Advisory Committee on Health Research, *Genomics and World Health*, Geneva: World Health Organization, 2002.

Aguda T. A., 'Human Rights and the Right to Development in Africa', *The Nigerian Institute of International Affairs*, Lecture Series No. 55, 1989.

Alston P., *Making and Breaking Human Rights: The UN's Specialised Agencies and Implementation of the International Covenant on Economic, Social and Cultural Rights*, Human Rights and Development Working Paper Series, London: Anti-Slavery Society, 1979.

Assistant Secretary for Environment, Safety, and Health, *Human Radiation Experiments Associated with the U.S. Department of Energy and Its Predecessors*, US Department of Energy, Washington DC, 20585, July 1995.

Behrman J. N., *Toward a New International Economic Order*, Paris: The Atlantic Institute for International Affairs, 1974.

Cook-Deegan R., Chan C., Johnson A., *World Survey of Funding for Genomic Research*, Stanford-in-Washington Program, Washington DC, 2000.

Declaration of Alma-Ata, International Conference on Primary Health Care, Alma-Ata, USSR, 6–12 September 1978.

Directive 98/44/EC on the Legal Protection of Biotechnological Inventions, of the European Parliament and of the Council of 6 July 1998.

Frank A. G., 'Rhetoric and Reality in the New International Economic Order' in Addo H. (ed.), Transforming the World Economy? Nine Critical Essays on the New International Economic Order, Hodder and Stoughton, London and Westview Press, Boulder, CO, in association with the United Nations University, Tokyo, 1985.

Hartlev M., *Biotechnology – A Challenge to Human Rights?*, 9th World Congress of Bioethics, Rijeka, Croatia, 6 September 2008.

Idris K., *Intellectual Property: A Power Tool for Economic Growth*, WIPO, 2003.

Intergovernmental Working Group on Public Health, *Draft Global Strategy and Plan of Action on Public Health, Innovation and Intellectual Property*, Geneva: World Health Organization, 2007.

International Indian Treaty Council, *Consolidated Indigenous Shadow Report – Examination of the United States 4th, 5th and 6th Periodic Reports of April 2007*, submitted to the Committee on the Elimination of Racial Discrimination, February 2008.

Kirby M., 'The Universal Declaration on the Human Genome – A Vital UNESCO Initiative', 2nd International Summit of National Bioethics Advisory Commissions, Tokyo, Japan, 1998.

Nuffield Council on Bioethics, *The Ethics of Patenting DNA – A Discussion Paper*, Nuffield Council on Bioethics, London, 2002.

Oldham P., 'Global Status and Trends in Intellectual Property Claims: Genomics, Proteomics and Biotechnology', submitted as UNEP/CBD/WG-ABS/3/INF/4 by the European Community to the Third Meeting of the Ad-Hoc Open-Ended Working Group on Access and Benefit-Sharing, United Nations Convention on Biological Diversity, Bangkok-Thailand, 14–18 February 2005.

Proposal by Argentina and Brazil for the Establishment of a Development Agenda for WIPO, World Intellectual Property Organization, 2nd Session, Geneva, 26–30 June 2006.

Proposal by the United States of America for the Establishment of a Partnership Program in WIPO, Inter-sessional Intergovernmental Meeting On a Development Agenda for WIPO 1st Session, Geneva, 11–13 April 2005, IIM/1/2, 18 March 2005.

Report of the Committee of Inquiry into Human Fertilisation and Embryology, Cm 9314 (Warnock Report), London: HMSO, 1984.

Roco M. C., Bainbridge W. S. (eds), *Converging Technologies for Improving Human Performance: nanotechnology, biotechnology, information technology and cognitive science*, Arlington, Virginia: National Science Foundation, 2002.

Sanu O. E., *The Lomé Convention and the New International Economic Order*, The Nigerian Institute of International Affairs, Lecture Series No. 18, 1979.

Statement by the Vice-Chairman of Sub-Committee I on (Item 2) of the Work Programme: 'Status, Scope, Functions and Powers of the International Machinery' at the 55th Meeting held on 4 August 1972 (A/AC.138/SC.I/L.17), in *DOA Law of the Sea*, p. 235.

Swenarchuk M., *The Harvard Mouse and All That: Life Patents In Canada*, CELA Publication #454, Toronto: Canadian Environmental Law Association, 2003.

Trials of War Criminals before the Nuremberg Military Tribunals under Control Council Law No. 10. Nuremberg, October 1946–April 1949, Washington, DC: US GPO, 1949–1953.

Turković K., 'Informed Consent: Facing Challenges of Multiculturalism', plenary talk, 9th World Congress of Bioethics, Rijeka, Croatia, 6 September 2008.

United Nations Commission on Human Rights, Press Releases, 57th Session, 25 April 2001.

United Nations Development Programme, *Making Global Trade Work for People*, pp. 221–2.

Online sources

American Geological Institute, 'Law of the Sea (2-2-05)', available online at http://www.agiweb.org/gap/legis109/lawofthesea.html (accessed 18 August 2006).

Annas G. J., 'Governing Biotechnology', *Global Agenda Magazine*, available online at http://www.globalagendamagazine.com/2006/Annas.asp (accessed 6 April 2006).

'Baby born after just 22 weeks leaves ward', *Guardian*, 20 February 2007, available online at http://www.guardian.co.uk/world/2007/feb/20/usa (accessed 25 August 2008).

Center for Strategic Economic Studies, *A Primer on the Knowledge Economy*, CSES: Victoria University, 2000, available online at http://www.cfses.com/documents/knowledgeeconprimer.pdf (accessed 16 April 2003).

Chapman C. (Minority Rights Group International), Statement at the 55th Session of the Sub-Commission on the Promotion and Protection of Human Rights, 11 August 2003, available online at http://www.unhchr.ch/huricane/huricane.nsf/0/670B54E219DC300DC1256D7F004FDC49?opendocument accessed (accessed 11 August 2008).

Chairman of the Legal Commission, 'Project of an International Instrument for the Protection of the Human Genome', Paris: UNESCO, available online at http://www.unesco.org/ibc/en/genome/juridique/instrument.html (accessed 24 November 2001).

CIOMS (Council for International Organizations of Medical Sciences), *International Guidelines for Ethical Review of Epidemiological Studies*, available online at http://www.cioms.ch/frame_1991_texts_of_guidelines.htm (accessed 11 January 2006).

COMEST (The World Commission on the Ethics of Scientific Knowledge and Technology), Paris: UNESCO, available online at http://portal.unesco.org/shs (accessed 21 May 2006).

Conference on Financing for Development (FfD) – ELDIS: The Gateway to Development Information, 22 March 2002, available online at http://www.eldis.org/static/DOC1942.htm (accessed 12 November 2006).

Convention for the Protection of Human Rights and Dignity of the Human Being with regard to the Application of Biology and Medicine, CETS No. 168, Council of Europe, available online at http://conventions.coe.int (accessed 29 March 2006).

Espiell, H. G., *Methodology for the Preparation of an International Instrument for the Protection of the Human Genome*, Paris: UNESCO, 1994, available online at http://www.unesco.org/ibc/en/genome/juridique/methodologie.html (accessed 24 November 2001).

European Parliament Resolution on the Patenting of BRCA1 and BRCA2 ('Breast Cancer') genes, Official Journal C 87E, 11 April 2002, p. 265, available online at http://europa.eu.int/eur-lex/pri/en/oj/dat/2002/ce087/ce08720020411en02650266.pdf

Fagan A., 'Human Rights' in Fieser J., Dowden B. (eds) *The Internet Encyclopaedia of Philosophy*, available online at http://www.iep.utm.edu/ (accessed 29 September 2006).

'First UN Development Decade', available online at http://www.nationsencyclopedia.com/United-Nations/Economic-and-Social-Development-FIRST-UN-DEVELOPMENT-DECADE.html (accessed 29 August 2006).

Fidler D. P., *International Law – Global Public Goods for Health – A Reading Companion*, Geneva: World Health Organization, available online at http://www.who.int/trade/distance_learning/gpgh/en/index.html (accessed 2 February 2008).

Fortun, M. A. 'Celera Genomics: The Race for the Human Genome Sequence' in *Encyclopedia of Life Sciences*, John Wiley & Sons, Inc., 2006, available online at http://mrw.interscience.wiley.com/emrw/9780470015902/els/article/a0005182/current/abstract (accessed 12 March 2008).

Geneva Declaration on WIPO, Summary, available online at http://www.cptech.org/ip/wipo/da.html (accessed 12 November 2006).

Hohfeld W., 'Fundamental Legal Conceptions' in Cook, W. (ed.), Fundamental Legal Conceptions as Applied in Judicial Reasoning: Essays by Wesley Newcomb Hohfeld, New Haven: Yale University Press, 1923. Also in 'Rights' in *The Stanford Encyclopedia of Philosophy*, available online at http://plato.stanford.edu/entries/rights/#2.1 (accessed 29 September 2006).

Human Genome Project Information [website], 'Ethical, Legal and Social Issues', available online at http://www.ornl.gov/sci/techresources/Human_Genome/elsi/elsi.shtml (accessed 12 April, 2006).

IBM Archives, available online at http://www-03.ibm.com/ibm/history/ (accessed 11 April 2006).

Jean G. V., 'Space Programs Poised for Major Expansion in Asia, November 2008', *National Defense Magazine*, available online at http://www.nationaldefensemagazine.org/archive/2008/November/Pages/SpaceProgramsPoisedforMajorExpansioninAsia.aspx (accessed 23 October 2008).

Marks J., 'Human Genome Diversity Project: Impact on Indigenous Communities', in Cooper, D. (ed.), *Encyclopedia of the Human Genome* (5 vols), London: Macmillan, available online at http://personal.uncc.edu/jmarks/pubs/main.html (accessed 12 April 2006).

Marshall E., 'Bermuda Rules: Community Spirit, with Teeth', *Science* 291 (5507), 2001, 1192, available online at http://www.sciencemag.org/cgi/content/full/291/5507/1192 (accessed 12 September 2006).

McKusick V. A., *The Human Genome Organisation, History, Purposes and Membership*, available online at http://www.hugo-international.org/abt_history.php (accessed 26 July 2008).

Morgan M., Chief Executive of the Wellcome Trust Genome Campus, *The Future is Genomic*, Wellcome Trust Sanger Institute, available online at http://www.sanger.ac.uk/HGP/draft2000/future.shtml (accessed 8 April 2006).

The National Commission for the Protection of Human Subjects of Biomedical and Behavioural Research, *The Belmont Report: Ethical Principles and Guidelines for the Protection of Human Subjects of Research*, 1979, Department of Health, Education, and Welfare, USA, available online at http://ohsr.od.nih.gov/guidelines/belmont.html#top (accessed 11 April 2006).

Office of the Commission on Human Rights, 'Human Rights in Development: The Content of the Right to Development', available online at http://www.unhchr.ch/development/right-02.html (accessed 29 August 2006).

—— Proclamation of Tehran, available online at http://www.unhchr.ch/html/menu3/b/b_tehern.htm (accessed 29 August 2006).

Office of the Press Secretary, The White House, 'A Renewed Spirit of Discovery: President Bush Announces New Vision for Space Exploration Program', 14 January 2004, available online at http://www.whitehouse.gov/news/releases/2004/01/print/20040114–11.html (accessed 18 August 2006).

Pardo A., 'The Origins of the 1967 Malta Initiative', *International Insights* 9 (2), 2003, 65–9, available online at http://website.lineone.net/~romweb/par_a01.htm (accessed 28 June 2006).

'Religion and Ethics-Taoism', BBC, available online at http://www.bbc.co.uk/religion/religions/taoism/taoethics/ethics_1.shtml (accessed 29 September 2006); 'The Mystical Non-Action', available online at http://jaan.kaplinski.com/philosophy/action.html (accessed 29 September 2006).

Robinson M., 'Constructing an International Financial, Trade and Development Architecture: The Human Rights Dimension, Zurich', 1 July 1999, available online at http://www.unhchr.org

Sandalow D. B., 'Law of the Convention: Should the U.S. join?', *Policy Brief 137, The Brookings Institution*, 2004, available online at http://www.brook.edu/comm/policybriefs/pb137.htm (accessed 18 August, 2006).

'Scientists crack human code', BBC, Monday 26 June 2000, available online at http://news.bbc.co.uk/1/hi/sci/tech/805803.stm (accessed 6 June 2007).

Segal Y., *The Israel Academy of Sciences and Humanities*, available online at http://www.daat.ac.il/daat/kitveyet/assia_english/segal.htm (accessed 7 May 2006).

Skyrme D. J., 'The Global Knowledge Economy: and its implication for markets', *Insight* 21, 1997, available online at http://www.skyrme.com/insights/21gke.htm (accessed 10 November 2006).

The Genographic Project [website], available online at https://www3.nationalgeographic.com/genographic/faqs_about.html#Q1 (accessed 12 April 2006).

The Human Genome Diversity Project, available online at, www.stanford.edu/group/morrinst/hgdp.html (accessed 12 April 2006).

The Human Genome Program of the US Department of Energy Office of Science, Summary of Principles Agreed at the First International Strategy Meeting on Human Genome Sequencing, Bermuda, 25–28 February 1996, available online at http://www.ornl.gov/sci/techresources/Human_Genome/research/bermuda.shtml#1

Treaty to Protect Genetic Commons, available online at http://sbb.collectifs.net/Textes/Rifkin.htm (accessed 13 October 2006).

UNESCO, *Proceedings of the First Session of the IBC*, Paris: UNESCO, 1993, available online at http://portal.unesco.org/shs (accessed 15 May 2006).

—— *Proceedings of the Second Session of the IBC*, Paris: UNESCO, 1994, available online at http://portal.unesco.org/shs (accessed 15 May 2006).

—— *Report of the First Meeting of the Legal Commission of the IBC*, Paris: UNESCO, 1994, available online at http://www.unesco.org/ibc/en/genome/juridique/r1.html (accessed 24 November 2001).

—— *Report of the Second Meeting of the Legal Commission of the IBC*, Paris: UNESCO, 1994, available online at http://www.unesco.org/ibc/en/genome/juridique/r1.html (accessed 24 November 2001).

—— *Report of the Third Meeting of the Legal Commission of the IBC*, Paris: UNESCO, 1994, available online at http://www.unesco.org/ibc/en/genome/juridique/r1.html (accessed 24 November 2001).

—— *Proceedings of the Third Session of the IBC*, Paris: UNESCO, 1995, available online at http://portal.unesco.org/shs (accessed 15 May 2006).

—— *Report of the Fourth Meeting of the Legal Commission of the IBC*, Paris: UNESCO, 1995, available online at http://www.unesco.org/ibc/en/genome/juridique/r1.html (accessed 24 November 2001).

—— *Report of the Fifth Meeting of the Legal Commission of the IBC*, Paris: UNESCO, 1995, available online at http://www.unesco.org/ibc/en/genome/juridique/r1.html (accessed 24 November 2001).

—— *Proceedings of the Fourth Session of the IBC*, Paris: UNESCO, 1996, available online at http://portal.unesco.org/shs (accessed 15 May 2006).

—— *Report of the Sixth Meeting of the Legal Commission of the IBC*, Paris: UNESCO, 1996, available online at http://www.unesco.org/ibc/en/genome/juridique/r1.html (accessed 24 November 2001).

—— *Report of the Seventh Meeting of the Legal Commission of the IBC*, Paris: UNESCO, 1996, available online at http://www.unesco.org/ibc/en/genome/juridique/r1.html (accessed 24 November 2001).

—— *Report of the Eighth Meeting of the Legal Commission of the IBC*, Paris: UNESCO, 1996, available online at http://www.unesco.org/ibc/en/genome/juridique/r1.html (accessed 24 November 2001).

—— 'Avicenna Prize for Ethics in Science', available online at http://portal.unesco.org/shs (accessed 21 May 2006).

—— 'Ethics', available online at www.unesco.org/ethics (accessed 21 May 2006).

—— 'International Bioethics Committee', available online at http://portal.unesco.org/shs (accessed 21 May 2006).

—— 'Meetings of the Legal Commission', available online at http://www.unesco.org/ibc/en/genome/juridique/r1.html (accessed 24 November 2001).

United Nations, 'Sub-Commission Adopts Resolutions on Right to Return, Human Rights Conference', Press Release, 54th Session, 15 August 2002, available online at http://www.unhchr.ch/huricane/huricane.nsf/0/31071122AE996F1BC1256C17002739D2?opendocument (accessed 9 August 2008).

United Nations Conference on Trade and Development (UNCTAD), Annotated Agenda of the Tenth Session of the United Nations Conference on Trade and Development, 12 February 2000, available online at http://www.unctad.org/en/docs/ux_td379.en.pdf (accessed 21 October 2003).

—— 'UNCTAD XI Concludes with Adoption of 'Sao Paulo Consensus' Focusing on Poverty Reduction in Least Developed Countries', Press Release, 21 June 2004, available online at http://www.unis.unvienna.org/unis/pressrels/2004/tad1994.html (accessed 12 November 2006).

US Department of Energy Office of Science, Office of Biological and Environmental Research, Human Genome Program, 'About the Human Genome Project', available online at http://www.ornl.gov/sci/techresources/Human_Genome/project/about.shtml (accessed 18 June 2008).

US National Institutes of Health (NIH), 'Directives for Human Experimentation: Nuremberg Code', available online at http://www.nihtraining.com/ohsrsite/guidelines/nuremberg.html (accessed 5 April 2006).

Wellcome Trust, Sharing Data from Large-Scale Biological Research Projects: A System of Tripartite Responsibility, 2003, available online at www.wellcome.ac.uk/assets/wtd003207.pdf

Wild, J., 'The Future for Patents on Life', *IP Matters*, 2002, available online at http://scientific.thomson.com/free/ipmatters/patlife/8180006/ (accessed 13 October 2006).

Winston M., *On the Indivisibility and Interdependence of Human Rights*, paper presented at the 20th World Congress of Philosophy, Boston, 10–15 August 1998, available online at http://www.bu.edu/wcp/Papers/Huma/HumaWins.htm (accessed 20 January, 2006).

World Health Organization, 'WHO reports 10 million TB patients successfully treated under "DOTS" 10 years after declaring TB a Global Emergency', available online at http://www.who.int/mediacentre/news/releases/2003/pr25/en/ (accessed 12 January 2005).

World Intellectual Property Organization, Inter-sessional Intergovernmental Meeting on a Development Agenda for WIPO, 1st Session, Geneva, 11–13 April, 2005, Proposal by the United Kingdom, IIM/1/5, 2005, available online at http://www.wipo.int/edocs/mdocs/mdocs/en/iim_1/iim_1_5.doc.

—— 'WIPO Member States Agree on a Way Forward for Discussions on a WIPO Development Agenda', Press Release 461, Geneva, 2 October 2006, available online at http://www.wipo.int/edocs/prdocs/en/2006/wipo_pr_2006_461.html (accessed 28 October 2006).

World Medical Association, Declaration on the Human Genome Project, 1992, adopted by the 44th World Medical Assembly, Marbella, Spain, September 1992 and available online at http://www1.umn.edu/humanrts/instree/humangenome.html (accessed 3 July 2006).

Index